i

Silent Spring-Deadly Autumn of the Vietnam War

By

Patrick Hogan

Lived, researched and written by Patrick Hogan

Copyright © 2019 Patrick Hogan

For permission requests, please contact: ssdamail@ptd.net

ISBN: 978-1-7325474-1-4

April 2019

Editing by KIRKUS

Publisher:

Whatnot Enterprises, LLC
Ephrata, PA 17522

Website:

https://www.ssdavw.com

This book and blessing is for all my brothers and sisters who served the United States in South Vietnam and to all veterans of foreign wars, no matter where they may have fought.

May the Lord bless you and keep you;
May the Lord make his face to shine upon
you and be gracious to you;

May the Lord lift up his countenance upon
you and give you peace.

(Numbers 6:24-26)

The last trip home no one wanted to take.

We all gave something of ourselves in Vietnam,
but to many gave their all.

A Special Thanks and Personal Note

First and foremost, I thank God for moving and helping me to complete this work. Next, I especially thank my loving wife, Georgia, for understanding and putting up with me for the last half-a-century.

I thank my wonderful children and their spouses: Tara, Scott & Cheryl, Kelly & Michael, and Dena & Cory for their loving support, and all my family and friends for their patience and understanding of why I spent so much time in front of my computer researching and writing this book.

Lastly, I would like to speak, by way of this book, to my great-grandchildren and those yet to be born. It is my heartfelt hope that one day you will have a chance to read this book and understand more about the war in Vietnam and how it affected my life and how it could even have affected your life and that of your future children.

I wish I could tell you that the pesticides I was exposed to in Vietnam so many years ago were not going to impact your health or your lives, but the real fact is: they very well might. Unfortunately, as I'm putting the finish touches to this book, no one really knows how my exposures will affect you or how they might impact your children's children because of the genotoxicity of all the unstudied hazardous chemicals used during the Vietnam War.

I hope and pray that the pesticides I was exposed to won't impact your life or the lives of your children. God willing, one day you will finally know how you and your health were affected by the Vietnam War, even though you were never there. I know it must be incomprehensible on many levels, but it's true nonetheless.

While I have dedicated this work to the men and women who served the U.S. in South Vietnam and other foreign wars, I wrote it for you and the future generations to come after you. In all probability, by the time you are old enough to read this book, I won't be around to explain to you what happened in Vietnam. However, I have confidence what I have written will.

As you continue reading, if you remember nothing else, remember this crucial lesson—one that took me a long time to realize: *actions speak louder than words*. Words like love, honor, decency, and integrity are simple words. They have no meaning until you give them worth by your actions. So, when it comes to issues of what's right or what's wrong, there is no middle or neutral ground; or to put it more eloquently, as the British philosopher John Stuart Mill said in his 1867 speech:

> "Let not any one pacify his conscience by the delusion that he can do no harm if he takes no part, and forms no opinion. Corrupt men need nothing more to accomplish their deeds than that good men should look on and do nothing."

Table of Contents

Glossary of Acronyms and Abbreviations Used

µg = Microgram

µg/m² = Micrograms per square meter

2, 4, 5-T = 2, 4, 5-Trichlorophenoxyacetic Acid

2, 4,-D = 2, 4-Dichlorophenoxyacetic Acid

BTEX = Benzene, Toluene, Ethylbenzene, and Xylenes

CDC = Centers for Disease Control

CDD = Chlorinated Dibenzo-p-Dioxins

COI = Chemicals of Interest

CONARC = Continental Army Command

COPD = Chronic Obstructive Pulmonary Disease

db = decibel

DDT = Dichloro-diphenyl-tricloroethane

DLC = Dioxin Like Compounds

DOD = Department of Defense

DVA = Department of Veterans Affairs

EPA = Environmental Protection Agency

FBGL = Fasting Blood Glucose Level

FIGMO = Freak-it I Got My Orders

GED = General Equivalency Diploma

GERD = Gastroesophageal Reflux Disease

HCB = Hexachlorobezene

HDL = High-Density Lipoproteins

ICMESA = Industrie Chimiche Meda Societa Azionaria

JP-4 = Jet Fuel

kg = kilogram

KP = Kitchen Patrol

Klick = 0.6 Miles

L = Liter

LD_{50} = 50 percent lethal dosage

LDL = Low-Density Lipoproteins

LES = Lower Esophageal Sphincter

MACV = Military Assistance Command, Vietnam

MCL = Maximum Contaminant Level

mg/L = milligrams per liter

MOS = Military Occupational Specialty

MRE = Material Release Expeditor

MRO = Material Release Order

MRSA = Methicillin-Resistant *Staphylococcus aureus*

NAM = The National Academy of Medicine

NAS = National Academy of Science

NIH = National Institutes of Health

NRC = National Research Council

NVA = North Vietnamese Army

OSS-TMP = O,S,S-Trimethyl Phosphorodithioate

PAHs = Polycyclic Aromatic Hydrocarbons

PCB = Polychlorinated Biphenyls

PCP = Pentachlorophenol

PCP = Primary Care Physician

PG = Pyoderma Gangrenosum

PLL = Prescribed Load Listing

ppb = parts per billion

ppm = parts per million

PROV = Provincial

PSP = Pierced Steel Planking

PTSD = Post Traumatic Stress Disorder

PX = Post Exchange

R&R = Rest & Relaxation

REFRAD = Release From Active Duty

ROKS = Republic of South Korea Marines

TCBQ = Tetrachlorobenzoquinone

TCDD = 2,3,7,8-Tetrachlorodibenzodioxin

TEF = Toxic Equivalency Factor

TIBA = Triiodobenzoic Acid

TO&E = Table of Organization and Equipment

UC = Ulcerative Colitis

USARV = United States Army Vietnam

VC = Viet Cong

VOC = Volatile Organic Compounds

Chapter 1
View from Under the Bus

With so many pressing issues and difficulties in today's world, some view the continuing health problems and illnesses surrounding "in-country" (boots-on-the-ground) Vietnam veterans as inconsequential in comparison.[1] But make no mistake about it: even though President Nixon's magnificent political declaration of "peace with honor" and the fall of Saigon in April of 1975 were supposed to symbolize the ending of the war in Vietnam, it hasn't ended for me or for other military personnel who served in South Vietnam.

The last battles of the Vietnam War are still being waged by veterans all over the country. Even today, as you read this book, the skirmishes over the pesticides we veterans were exposed to and the presumed illnesses they caused are still raging. Our battles are not with the Viet Cong (VC) or the North Vietnamese army. Instead, our conflicts are with the myriad cancers, illnesses, and health issues that we—and even many of our children—must battle with, day in and day out.[2]

Our clashes and struggles are with the bureaucratic systems of the Department of Veterans Affairs (DVA) and with the very government that sent us to South Vietnam in the first place. You can rest assured the Vietnam War isn't over—not by a long shot.

The fact that veterans who served in Vietnam must struggle with innumerable cancers, illnesses, and other health problems almost certainly caused by the highly toxic synthetic chemicals they were exposed to during the war is shameful—especially when you consider that those contaminated pesticides were atomized into a fine mist and sprayed onto them by their government/military leaders. More ominously, the substances veterans were exposed to were long-enduring herbicides and insecticides, still quite capable of claiming new lives daily—even though the Vietnam War has been diplomatically resolved for half a century.

As if being unprotected from health-damaging pesticides wasn't bad enough, boots-on-the-ground veterans were tactlessly thrown under the bus, by those very same government officials. Then, to add insult to injury, after being treacherously betrayed by administrative bureaucrats—who, for the most part, were sitting safely in Washington, DC, during the war— they proceeded to back their bus up and run over veterans again and again, just for good measure. This book is my view from under that governmental bus.

By the way, if you feel my use of the phrase *treacherously betrayed* was too harsh after you finish reading this book, let me know what your thoughts are. Was my comment too robust or just right—or do you think it's a little on the weak side?

Agent Orange and Its Deadly Companions

Without a doubt, the term *Agent Orange* will be remembered in dishonor, no matter what the US government may assert. Still, for most people living in today's world, the term doesn't have much significance other than that it had something to do with "that war in Asia." However, for the military personnel who served in South Vietnam, Agent Orange is synonymous with illness, pain, suffering, and death.

Official US government archives contain the files of more than 58,220 US service personnel who died in Vietnam. They were considered the ultimate casualties of that war. In addition, scores more—over three hundred thousand[3]—were recorded as injured or maimed. Disappointingly, not chronicled in those sobering statistics are the hundreds of thousands of soldiers, marines, and sailors who were injured or killed by the chemical pesticides sprayed on them during the war but didn't understand it until many years later.

Ever since the official end of the Vietnam War, the US government, the DVA, and the Department of Defense (DOD) have denied, obstructed, and rebuffed almost all attempts at etiologically or medically linking any illness or disorder with exposure to Agent Orange or to any of the other less-publicized—but just as deadly—pesticides deployed during the war. Our government and DVA continue their denials and obstructive actions with policies and procedures for which their mantra could very well be "Delay and deny until they all die."[4]

Dr. Jeanne Stellman, an epidemiologist who has spent decades studying Agent Orange for the American Legion and the National Academy of Sciences, said the following in a 2009 *Chicago Tribune* interview with Jason Grotto and Tim Jones:

> We do not know the answer to the question: What happened to Vietnam veterans? The government doesn't want to study this because of international liability and issues surrounding chemical warfare. And they're going to win because they're bigger and everybody's getting old and there are new wars to worry about.[5]

What a sad commentary on our legislative and military leaders. It's shameful and disappointing that veterans must struggle with serious illnesses because of all the harmful pesticides they were exposed to while serving the United States. Dr. Stellman's observation of our leaders is an accurate, concise, and honest one. What our government has done over the years to Vietnam veterans is reprehensible. It would have been much better if our elected officials had chosen to bite the bullet and throw the chemical companies that made the extremely harmful pesticides and the officials who decided to use them under the bus instead of the veterans who were exposed to their nightmarish military concoctions.

At this point, you might be asking, "Just who are you to be talking about the US government and military in this way?" In reality, I'm just one of the many boots-on-the-ground Vietnam veterans trying to make sense of the overly complicated bureaucratic care-and-benefit system set up by our policymakers and the DVA to consider our exposure to Agent Orange and all the other intermingled destructive pesticides used during the Vietnam War.

3

Ironically, the US military test-sprayed the first of their many hazardous herbicides on my fifteenth birthday, during Operation Hades, over a small village north of Dak To, located in Kon Tum Province. This early creation was called Dinoxol, and five short years after that unfortunate day, I would find myself in Vietnam.

I was stationed in Vietnam for two years, nine months, and twenty-two days—from September 1966 through June 1969. During this time—as chronicled, thanks in part to work done by Dr. Jeanne Mager Stellman and her group—I was located in geographic areas of South Vietnam that had been sprayed directly with several tactical grade pesticides. So despite the fact, we saw personnel with backpacks spraying around the cantonment areas, and on occasion aerial spraying by helicopters or planes, we were never told exactly what was in those fine mists being spewed out on us.

All the same, according to recently declassified military records, we were showered with, at a minimum, the herbicides Agent Orange and Agent White and the insecticides malathion and DDT.

I'm sure by now almost everyone has heard about Agent Orange, perhaps even to the point of exasperation. But have you ever heard anything—anything at all—about the following health-damaging chemicals contained in the tactical pesticides used by the military during the war in South Vietnam?

**2,4-D (2,4-Dichlorophenoxyacetic acid): one of the active ingredients in Agent Orange (also called Dinoxol) & in Agent White (also called Tordon 101)

**2, 4, 5-T (2,4,5-Trichlorophenoxyacetic acid): one of the active ingredients in Agent Orange

**Picloram (4-amino-3,5,6-trichloro picolinic acid): one of the active ingredients in Agent White

**HCB (Hexachlorobenzene): A toxic side reaction compound found in Agent White. It has also been classified as a POP (Persistent Organic Pollutant) by the Stockholm Convention.

**Triisopropanolamine: one of the "inert" components in Agent White

4

**TCDD (2,3,7,8-tetrachlorodibenzo-para-dioxin, better known as dioxin): a deadly contaminant found in Agent Orange, Agent White and other pesticides

**DLC (Dioxin-like compound): toxic impurities found in Agent Orange, Agent White, malathion, and many other pesticides

**Benzene, toluene, ethylbenzene, and xylenes (BTEX): toxic substances contained in JP-4 and various petroleum products used to dilute oil base pesticides

**Malathion: an insecticide from the family of pesticides called organophosphates

**OSS-TMP (O,S,S-trimethyl phosphorodithioate): a more toxic storage contaminant of malathion

**Malaoxon: a more toxic metabolite[6] of malathion

I suspect that you've heard little or nothing about any of these hazardous chemicals being used during the Vietnam War or about the fact that almost everyone stationed there was exposed to them multiple times.

Now keep in mind that the preceding was the "short list" of chemicals. As you will soon learn, there were an endless number of other chemicals—both known and still unidentified—contained in the few tactical grade pesticides I have noted. All the same, while we were in Vietnam, we didn't know those toxic chemicals were contained in the pesticides being sprayed on us, let alone the harm they could do to our health, especially synergistically. I, like many others, wanted to believe our military leaders when they told us that the substances being sprayed on us were safe, or as various officials had allegedly said, "They are just as safe as your backyard bug spray or weed killer. In fact, you can shower in the stuff, and it wouldn't be a problem."

Even though we didn't know the harm these pesticides were capable of causing, the chemical companies and the chemists who worked for them, as well as our government scientists, knew full well what they were capable of doing to any human, animal, or plant unlucky enough to be put in their path. They knew the tremendously harmful and systemically damaging nature of these very quickly and cheaply made heavy-duty military pesticides.

5

Unfortunately, it is chiefly because chemical firms produced their tactical pesticides so quickly and cheaply—for the exclusive use of our military—that the finished products were significantly more dangerous. In fact, any person in a six-plus-mile range of any base or operational location being sprayed might have had their health compromised and wouldn't even realize it.

I hope most of you still remember the old Frank Robinson classic 1973 baseball quote, "Close don't count in baseball. Close only counts in horseshoes and hand grenades." Well, you can add herbicides and insecticides to the end of that astute assessment. The sad irony is that the military pesticides could have been made less toxic and detrimental if the chemical companies had crafted them far more slowly and methodically via a low-heat method. But the low-heat, extra-care techniques would have been substantially more expensive for the military and markedly less profitable for the chemical companies manufacturing them.

As a consequence, the clandestine pesticides and associated chemicals military personnel were exposed too were considerably more hazardous and health altering. To put it bluntly, our lifelong health and well-being would suffer as a result of exposure to these "classified" nightmarish herbicides and insecticides. Deplorably, the adverse health problems would surface years or decades after the war, as aging also played a role.

Unfortunately, the very same pesticides that are causing veterans' health problems are also quite capable of causing significant genetic injuries—chromosomal damage—which could be passed on and have detrimental health impacts on their children, their grandchildren, and, quite credibly, even their grandchildren's children.

Endnotes—Chapter 1

1. Postwar Shock Besets Veterans of Vietnam - New York Times August 20, 1972, by John Nordheimer

2. Agent Orange and Birth Defects - By Betty Mekdeci, Executive Director Birth Defect Research for Children

3. Vietnam War Facts, Stats and Myths - US Wings

4. VA to vets: 'Delay, deny, wait till they die'—By Craig M. Wax The Washington Times - December 1, 2015

5. Jason Grotto and Tim Jones, "Agent Orange's Lethal Legacy: For U.S., a Record of Neglect," *Chicago Tribune,* Dec. 4, 2009

6. Metabolites are the substances formed as part of the natural biochemical process in which a parent chemical is changed into different break-down compounds. For example, the insecticide malathion, once created and sprayed into our environment will start breaking down or changing its chemical structure to form monocarboxylic and dicarboxylic acid derivatives and malaoxon.

Chapter 2
How My Search Began

A broad ridge of hot, sticky, unseasonably humid air was stuck over my region of Pennsylvania one evening in May of 2012. The unusually steamy weather was, in some ways, eerily reminiscent of the tropical humidity of Vietnam. I'd just finished dinner and was settling in for a quiet night of watching TV in our family room. But as it turned out, God had different plans.

Like many veterans, for forty-three years, I'd successfully put the Vietnam War behind me, but on that one particular evening, everything literally changed in a flash as I began to watch my favorite evening news program. The show was replaying the remarks President Obama had made at the commemoration ceremony of the fiftieth anniversary of the Vietnam War. The following are excerpts of that speech:

> And one of the most painful chapters in our history was Vietnam—most particularly, how we treated our troops who served there. You were often blamed for a war you didn't start when you should have been commended for serving your country with valor. You were sometimes blamed for misdeeds of a few, when the honorable service of the many should have been praised. You came home and sometimes were denigrated, when you should have been celebrated. It was a national shame, a disgrace that should

have never happened. And that's why here today we resolve that it will not happen again. And so a central part of this fiftieth anniversary will be to tell your story as it should have been told all along. It's another chance to set the record straight. That's one more way we keep perfecting our union—setting the record straight. And it starts today. Because history will honor your service, and your names will join a story of service that stretches back two centuries. Let us tell the story of a generation of service members of every color, every creed, rich, poor, officer, and enlisted who served with just as much patriotism and honor as any before you. Let's never forget that most of those who served in Vietnam did so by choice. So many of you volunteered.[1]

As President Obama continued to speak, describing the war and what we had endured when we came home, I started to feel annoyed. To borrow a phrase, I felt like my feathers were being ruffled. The more he talked, the more irritated I became. All of a sudden, it felt like a barrier had broken in my mind. Old memories came flooding in nonstop—recollections of experiences I hadn't thought of in decades; flashes combined with bits and pieces of old remembrances ran through my mind over and over again. They were memories of the time I had spent in Vietnam and more recent recollections of the illnesses and health issues I'd attributed to Agent Orange over the years—most of which the DVA denied and which I, of course, couldn't substantiate at the time. The deluge of feelings and images kept roaring into my mind planting themselves firmly into my thoughts.

I started recalling memories of high school friends who had been killed in Vietnam and those who had died well after the war, such as my friend Larry White, who died in 2009. I began remembering what we'd endured after returning from the war. Some of my recollections were good—others, not so much. All night and into the next day, the memories—especially the ones concerning Larry and our mutual health issues—kept popping up until they became a torrent that was just impossible to ignore any longer. I finally decided that I had to do something, even though I wasn't quite sure exactly what that something was.

I began my fact-finding journey by gathering all my old medical files and requesting my military records from the National Personnel Records Center. While waiting for my files, I started to investigate and research Agent Orange but quickly learned that I'd been exposed to many other harmful pesticides and hazardous chemicals while in the army. It didn't take me long to discover information and documents that flat out astonished me. In a few short days of investigative research, I found a great deal of disturbing but useful information that gave me a good start on my excursion into the maze of bureaucratic red tape and procedures known as the DVA's disability rating and compensation system.

On June 1, 2012, I refiled my disability claims with the DVA, and so my journey into the uncharted, murky waters of the DVA's complicated system began anew. However, before I delve into all the troubling evidence I discovered, I think a little background information would be helpful—starting with my friend Larry and his struggles.

Larry's Story

Trying to fit Larry's life story—or that of any in-country Vietnam veteran—into a few short pages honestly does not, and cannot, do justice to their lives or their struggles. Nonetheless, the accounts of their lives need to be told. Their stories need to be told over and over again so that we don't forget why and how Larry and countless other veterans suffered and died because of their service to the United States in Vietnam. Nor should we forget how they were treated by US bureaucrats after the war.

You won't find Larry's life history in any newspaper article or government archives. You won't see his name engraved on the massive, highly polished black granite Vietnam Memorial Wall in Washington, DC, even though he and thousands of other veterans were tragic and very real casualties of that war.

During all the years I knew Larry, we never talked much about our experiences during the war. Still, every now and then, a circumstance or particular word would come up in our conversations with family and friends that would trigger our memories, and we'd start talking to each other in terms and words that only we could possibly understand. On more than one occasion, we garnered some strange looks and had to explain ourselves and our cryptic comments about "the Nam" to our friends or families.

Larry and I had been in South Vietnam during roughly the same time frame, but we had never actually met until my daughter, Dena, and his son, Corey, started dating. Still, just the mere fact that we'd been in the country at about the same time predetermined our common bond.

We shared not only a lifetime membership in the Agent Orange Club of Uncle Sam but also the Tet Offensive[2] of 1968. We shared many of the same life-altering and life-affirming experiences that service in South Vietnam offered. We had choked and coughed on the same smoke from burning dung and trash as well as on the dust, sand, and dirt kicked up into the air by the war and the movement of vehicles and men. We had been drenched by the same unsympathetic monsoon rains and had met up with many of the same inhospitable insects, snakes, and other unfriendly critters that called Vietnam home.

While we shared many of the same experiences Vietnam had to offer, I still consider myself to be one of the more fortunate veterans. So even though I have had a multitude of health problems over the years and have been in bad shape on more than a few occasions, I'm still kicking, although not quite as high as I used to. Even after my doctors removed and rearranged a few important body parts, I'm still alive and above ground. Yeah, I'm one of the lucky ones—so far, anyway. But there are literally tens of thousands who served in Vietnam—like Larry—who haven't been so fortunate.

Lawrence C. White was just an ordinary, fun-loving guy who enjoyed life and a really good party. Back in 1966, he was just another rambunctious eighteen-year-old kid who was fresh out of high school, full of promise and dreams with his whole life still ahead of him. Larry wanted more out of life than just a job. He wanted to go to college and do something with his life. But as circumstances would dictate, he just happened to be born at the right time (or the wrong time, depending on your view) to be drafted by the Selective Service System for the continually broadening Vietnam War.

Larry tried everything short of leaving the country and going to Canada to get out of the draft; after all, it wouldn't have been a great hardship to run off to Canada. At worst, it would only have been a minor inconvenience. In any event, all his heroic efforts were for naught. Uncle Sam still wanted him for the war—even with all his feigned illnesses and contrived problems. So despite all his protesting and fighting against being drafted, he chose, in the end, to serve his country, and he accepted being drafted. He didn't seek a college deferment or rush off to get married and have children. He didn't slink away or buy a bus ticket to Canada, as so many others had done. He served his country, albeit reluctantly.

The area around the base where Larry was ultimately stationed, like most of South Vietnam, was a lush tropical jungle—a heavily canopied sanctuary from which Viet Cong insurgents would attack, only to retreat afterward and disappear back into the thick vegetation as quickly as they'd appeared. Unknown to Larry, to myself, and to the hundreds of thousands of other military personnel was the fact that our future health and fates were being ever so carefully crafted and strategically planned by our government and its military establishment. We were being purposefully placed in the path of our government's and military's classified covert pesticide spraying programs. These deliberate but untried strategies would, in time, become known by militaristic code names, such as Operation Hades, Project Pink Rose, Operation Ranch Hand, Operation Trail Dust, and Operation Fly Swatter. However, no matter what the operational name was, it would be the aerial spraying of Agent Orange, Agent White, DDT, malathion, and a host of other color-coded, nightmarish, "sunset" herbicides and insecticides that would be problematic and life altering for most, if not all, of the military personnel stationed in Vietnam. The ever-expanding airborne dispersal of so many harmful speedily produced pesticides would create a real-life quagmire of illnesses and problems for the people exposed to them.

Ultimately, it would be our exposure to a multitude of organic chemicals that would produce the invisible cellular and genetic trauma that would unknowingly come home with us from Vietnam and relentlessly pursue us for the rest of our lives.

We all expected to be in danger from the enemy in Vietnam, but never in our wildest imaginings did we conceive of being placed in harm's way by our own military's atomized spraying of tens of millions of gallons of extremely harmful herbicides and insecticides. These profoundly dangerous chemicals would, over time, turn out to be some of the most prolific killers and cripplers of the Vietnam War—destructive adversaries far more malevolent than the enemy we were fighting.

The strategies developed and implemented by the United States to destroy insects and millions of acres of South Vietnam's vast jungles have left an appalling legacy for the personnel who served there, whether or not the government is willing to admit it. All the same, Larry and I were

13

only two of the roughly three million service personnel deployed to South Vietnam who were sprayed with those harmful agents, which means there are still a lot of untold stories about the Vietnam War and its chemical aftermath.

Larry served his mandatory tour of duty in South Vietnam, and after he returned home, like many of us, all he wanted to do was leave the miserable war behind and get on with his life. But inevitably, the war and its pesticides followed him home. They waited silently, hidden inside his body until they were ready to pounce and attack. They hung around patiently, advancing little by little and ever so slowly, for almost a decade before they sprang into action. They finally ambushed Larry for the first time when he was twenty-nine years old; just as his fledgling life was getting back to normal, he had a devastating heart attack—the first of many.

This first heart attack—diagnosed as ischemic heart disease—hit Larry pretty hard and almost killed him. Thanks to God, a lot of praying, and the skill of his medical team, he didn't die, although at the time, his doctor did think he might. All the same, Larry was left weak, traumatized, and unable to perform even the simplest of tasks without assistance. He was utterly unable to exert himself, let alone work. It took Larry a long time to bounce back from that first disastrous attack. Eventually, over time, he did recover, but his heart was ruthlessly damaged. So since Larry was completely unable to work, his wife, Lois, had to rejoin the workforce to make ends meet.

Larry spent his whole shortened life in and out of hospitals and doctors' offices. He struggled for his very existence while having to fight with the DVA for assistance. He lived the rest of his abbreviated life hand to mouth on what little help his family could provide. Eventually, he was aided by Social Security disability and state-sponsored social support programs. Essentially, welfare and food stamps became his family's means of subsistence.

Larry labored for decades to get help from the DVA. He filed several disability claims and had many compensation and pension exams with the organization, arguing all the while that his ischemic heart disease was a service-connected issue and a result of his tour in Vietnam and Agent Orange exposure. However, his statements and claims fell on deaf ears. No one at the DVA wanted to listen because his illness was not presumed.

His disability claims were repeatedly denied by the DVA because, in their opinion, there was "insufficient evidence" that his condition was service related. According to the government, there was just no substantiation that his ischemic heart disease was the result of his exposure to Agent Orange and his service in Vietnam. He was repeatedly told that none of the pesticides the government had used there could possibly have been the problem or the cause of his heart disease. The DVA officials told him it was more likely that his genetic makeup, and not his service in South Vietnam, was responsible for his medical problems.

As the years ticked by, Larry had several more major heart attacks and other health problems. In fact, his heart was so mercilessly damaged that eventually it was pumping at a very feeble volume. His actual heart ejection factor was lower than 35 percent. Ultimately, he had to have a defibrillator-pacemaker implanted in his chest to keep him alive. In the end, it wasn't his damaged heart that finally took his life. It was leukemia.

Appallingly, from the age of twenty-nine until his final battles with heart disease and leukemia on December 26, 2009, at the age of sixty-two, Larry never once received any assistance from the Department of Veterans Affairs. Disgracefully, it wasn't until October 30, 2010, that the DVA began readjudicating his claims—almost a year after his death. Ultimately, our government and the DVA were forced to concede that his ischemic heart condition and subsequent leukemia were related to his service in Vietnam and caused by the pesticides to which he was exposed. Nevertheless, this capitulation came way too late. Larry had lived his whole life in almost abject poverty and was barely able to make ends meet, with absolutely no assistance from the DVA.

Now I ask you, is that any way for our government to treat those they claim—at least in magnificent speeches—to be war heroes? Actions really do speak louder than words. The actions of our government officials and DVA speak volumes to me—even though they are mostly adversarial in nature.

Larry's story is only one account. Countless very similar and even more heart-wrenching life stories are out there and need to be told and listened to—stories of soldiers, marines, and sailors who can all relate and trace their illnesses, disorders, and health problems back to their service to the United States.

Larry served his required two years in the army with honor and dignity. One of those two years he spent in the hot, steamy jungles of South Vietnam. Thus, Larry became just one of the countless in-country veterans whose fate had been sealed in Vietnam. He was but one soldier who was injured and eventually killed by the Vietnam War, even though he died in Denver, Pennsylvania, some forty years after he had left Vietnam. Rest in peace, my old friend.

The Draft, 1965

Back in June 1965, I was almost eighteen, and like Larry, I had to register with the Selective Service System for the draft. After doing some inquiring, I discovered that in January of 1965, about 5,400 men had been drafted into the military. So with the Vietnam War continuing to escalate—and looming larger and larger on the horizon—I was pretty confident that I'd be drafted in short order. After discussions with my family and friends and several intense mental debates, I finally decided that my best course of action was to enlist. I believed it would be better for me to join rather than to hang around waiting to be drafted. I eventually decided to enlist in the regular army for two excellent reasons—or at least I thought they were brilliant at the time.

The first reason was the loyalty and obligation I felt toward God and our country, a "for the love of God, Mom, apple pie, and the flag" way of thinking—or perhaps it was just a symptom of my watching way too many cowboy and World War II movies.

The second reason was to endeavor to acquire some specialized training in the military—an education that I'd be able to use afterward as a civilian. And so in August 1965 I enlisted, just a few days after my eighteenth birthday. As had been arranged with my recruiter, I was assigned to Fort Dix, New Jersey, for basic combat training. Everything went as I had expected during basic until mid-October 1965. I was almost halfway through my training when I was hospitalized for five days with a severe upper respiratory infection. I came close to being "recycled." In fact, if my doctor had kept me in the hospital just one more day, I would have had to start basic training from the beginning. As it turned out, in retrospect, maybe I should have started over.

Despite my upper respiratory problem and the five training days lost, I still managed to finish in the top 10 percent of the class. After graduation, I was shipped off to the Army Security Agency at Fort Devens, Massachusetts, for advanced training. After a short time, my upper respiratory infection really took off, and between November 1965 and April 1966, I was hospitalized three times for a total of forty days.

My first two hospital stays were the result of viral pneumonia, which damaged my lungs and caused pleural thickening, fibrosis, and chronic blunting of my lung angles. The third hospitalization was for acute bronchitis. I don't recall much about my time at Devens, other than being in and out of the hospital, having severe chest pains, coughing, trouble breathing, and feeling sick most of the time. While my time at Devens was short, my illnesses made it feel much longer. Eventually, after my recovery from bouts of pneumonia and bronchitis, I was transferred in September 1966 to the 423rd Supply Company.

Shortly after settling into my new home base in Fort Lee, Virginia, our entire unit was notified that we were going to be deployed to South Vietnam, and everyone would be given a short leave. As you would expect, we all checked out with our company clerk and promptly left Fort Lee and headed for home. For me, it was a long, quiet bus ride from Virginia to New Jersey, which gave me plenty of time to contemplate my upcoming all-expense-paid trip to Southeast Asia, compliments of Uncle Sam. With the Vietnam War being in the news all the time, it left plenty of room for me to speculate and let my overactive imagination run wild, especially after watching so many war movies. Being a naive nineteen-year-old didn't help either.

Most of us realized that going to Vietnam wasn't going to be a pleasure trip, but no matter how much I anticipated what Vietnam would be like, the reality of being there paled in comparison to any war movie or news report. It wasn't intangible any longer; it was as real as it gets. I would not just see and hear the war, but I would taste it and smell it. More importantly, I would feel and experience each and every enjoyable, nasty, or painful event Vietnam had to offer.

I had literally just walked in the front door of my parents' home and was saying hello to them when the phone rang. It was our company clerk. He told me as politely as he could that we were all being ordered back to base immediately. He went on to explain that our departure date had been moved up in accord with our new deployment orders, now classified "secret." Since my leave had been cut short, all I had time for was to say goodbye to my parents, and I was off on my way back to base.

It was a day or so before everyone reported back, and we got everything squared away for our little expedition. We left Fort Lee early in the morning on a somewhat frosty but sunny September day, crowded into the cargo hold of an unadorned Hercules C-130 cargo plane. We flew as a unit, with all our essential supplies and equipment stowed in the middle aisle of the cargo aircraft. Consequently, because our gear was taking up the center space, we were relegated to either side of the cargo, sitting and sleeping on the built-in, even-less-than-budget-class web seating. Cramped, uncomfortable, and cold, we lived in the confined payload area of that C-130 for a couple of days en route to our final destination. All I could think of at the time was what Vietnam would be like and where were we going to land.

Interior section of a C-130 Hercules

Our first pit stop was Edwards Air Force Base, California. I still remember looking out one of the porthole-type windows as we were getting ready to land. Sadly, all that could be seen were miles of very bleak, foreboding, and rugged desert. "An omen of things to come?" I wondered to myself. All I could see were miles and miles of sand, scrub brush, and mountainous terrain, devoid of anything that even looked alive. "Nah, that's impossible," I retorted to myself. Vietnam was a jungle, not a vast, sandy desert.

We landed at Edwards for refueling, a much-needed bathroom break, and hot food. Actually, we did have C rations available while we were in flight . . . if you want to call that food. We ate quickly at the air force mess, cleaned up as best we could, and were loaded back on our planes for the next leg of the journey. In no time at all, we were off to our next destination, Hawaii, and another pit stop. Then it was off to Guam, where we stopped again. Guam was followed by stopovers in Okinawa and Tokyo, Japan. Then finally, we were on the last leg of our slow, twisty journey to Southeast Asia.

We were on that plane, by my best reckoning, for about four days in total. We slept on and off, passing the time as best we could. No one did a lot of talking because of the loud, droning noise of the propellers filling the cold, cramped cargo area. It was all we heard for days. Even after I got off the aircraft, I could still hear and feel the drone and vibrations of those props for almost a week.

Endnotes—Chapter 2

1. The White House - Office of the Press Secretary-May 28, 2012

2. Tet offensives were a series of coordinated attacks on major military installations (Cam Ranh being one those bases) by insurgent forces during the war. These North Vietnamese attacks were timed to coincide with their Lunar New Year, which they called "Tet."

Chapter 3

South Vietnam 1966

Our first home in a sea of sand

On September 30, 1966, we were advised by our loadmaster that our landing was imminent. So we geared up, donning flak vests and steel helmets for our arrival. We were all issued M-14s with several extra ammo clips. We double-checked to make sure everything was securely fastened down for landing, and finally we secured ourselves and waited to land. I was armed, locked, loaded, and ready for any eventuality, with a heart pounding out of my chest and gushing adrenaline to spare.

The plane landed without incident, but my heart was still racing. As we waited to exit the plane, I had a reality check, and a chilling thought ran through my mind. This wasn't a John Wayne flick or an Audie Murphy movie; this was real, and there were people here who were going to try to kill me—and me them. As I was still considering that thought, the rear cargo door was lowered, and I turned my focus to exiting the plane with military precision, ready to fight. Just who we were to fight was a mystery.

The only people in the area were the air force base workers and flight crews, all calmly going about their routine tasks. Needless to say, we got some pretty strange looks as we came running out of the C-130s—yelling. I don't know where our secret orders had us slated to go, but I'm pretty sure it wasn't Cam Ranh. In any event, Cam Ranh was where we landed, and that was where we stayed. We would find out later from our executive officer (XO) that our classified orders had been suddenly changed in midflight, with no explanation.

After landing, we loaded our equipment and ourselves onto 2.5and 5-ton trucks for transport to our new home. As my adrenaline rush subsided and my heart slowed to normal, I exited the truck and saw my new residence, which was nothing more than a cluster of tents arranged atop a broad plateau. Our new cantonment area was surrounded by a sea of sand, with vast, dense jungles to the west and southeast. So much for premonitions!

It was a few days before the fog in my head began to clear, my ears stopped throbbing, and the physical sensation of vibrating eased up. As my body was starting to return to normal and the haze in my head wore off, I woke up early one morning, just before dawn, to the twitter of birds and chatter of monkeys coming from the heavily forested jungle area about half a klick away in military jargon—or a third of a mile for you civilians.

I recall thinking how beautiful the jungle looked in comparison to our bleak, sandy surroundings. Listening to the playful sounds of the birds and animals coming from the thick forest was a new experience. It would be just one of the many unanticipated experiences Vietnam would have to offer.

Although our company was one of the first units in the area, many more would soon be joining us. The 423rd was, at least on paper, now part of the 96th Supply and Service Battalion (Direct Support) of the 504th Field Depot of the 1st Logistical Command. When we first arrived at Cam Ranh, we were immediately incorporated into depot operations and placed in an accelerated training program. Our speedy enhanced training courses were designed to provide Cam Ranh Depot with the necessary workforce it needed to hasten the receipt and storage of critical supplies as well as to deliver on-the-job training for everyone assigned to mission-critical fields of depot operations.[1]

The 423rd, along with the many other units attached to the army's 1st Logistical Command, has been credited with participation in the following Vietnam defense counteroffensives operations (VDC-O):

VDC-O Phase II—July 1, 1966–May 31, 1967

VDC-O Phase III—June 1, 1967–January 29, 1968

Tet Offensive 1968[2]—January 30, 1968–April 1, 1968

VDC-O Phase IV—April 2, 1968–June 30, 1968

VDC-O Phase V—July 1, 1968–November 1, 1968

VDC-O Phase VI—November 2, 1968–February 22, 1969

Tet Offensive 1969—February 23, 1969–June 8, 1969

No military organizational unit was more essential to the expanding war effort and buildup of US forces in Vietnam than the 1st Logistical Command. Before 1965, the US military was being supplied by the army's Pacific Command through a small support group under the 9th Logistical Command operating out of Okinawa. By mid-1965, after 1st Log Command (as we commonly called it) was established in Vietnam, the first depot could process about seventy thousand tons of incoming supplies per month. Just one year later, toward the end of 1966, the newly improved operation could handle well over seven hundred thousand tons of supplies per month, and that's not considering critical supply items brought in by air or Operation Red Ball.[3]

We worked around the clock, twelve to fourteen hours a day. Rain or shine, day or night, healthy or sick, we worked. We helped stock the warehouses and even helped with building our future billets. In fact, in September 1968, our battalion was awarded a Meritorious Unit Commendation with an oak leaf cluster, for our exceptional work and steadfast commitment to the war effort from October 1966 to June 1967.

All in all, despite the hot, humid weather and the sea of sand surrounding us—which made us feel like we were living inside an oven— we made pretty good progress. After several months, we moved out of our tents and into our new no-frills metal-roofed one-story barracks, which we called "hooches." Even though our hooches were built to be more enduring, they were still crude camping-out-style structures. We did have electricity most of the time, but of course, we still had no air

23

conditioning or running water for showers or toilets. Instead, we used the outdoor latrines and showers that we brought over with us from our tent city location.

Constructing our "hooch" area

Our showers were modest wooden buildings with fifty-five-gallon metal drums attached to the roofs. (I can't say what had been in the drums first.) The steel drums were filled with water daily and heated by the sun, so once in a while, we did get a sun-warmed shower instead of a cold one. Of course, we always had the option we had when we first arrived—wait for rain and then shower. The sight of a bunch of naked men showering in the rain wasn't very pretty, but it worked. Even though our showers were crude, they were better than the alternative.

Our bathroom facilities were bigger versions of the modest, very smelly outhouse. While our showers had the water basin on the roof, the outdoor toilets had them under the seats, and they had been cut in half. The half drums essentially served as reservoirs for our daily deposits. While the smell was nasty, an even worse drawback was the fact that every day, the contents of those drums had to be dragged outside, mixed with a combination of gasoline and kerosene, and lit on fire.

Honey-bucket detail

Unfortunately, after you ignited the contents, the black smoke would billow up almost immediately from what we jokingly referred to as "honey buckets." It's hard to describe the unpleasant, pungent aroma of the dark smoke wafting up from the burning combination of dung and urine blended with gasoline and kerosene. The clingy choking smoke and the overpowering smell hovered in the humid air and, at times, even enveloped you. When it did, the stench would remain on you and your fatigues for some time after you finished the detail. While I hate to admit it, I was unlucky enough to be assigned to that detail on a few occasions. All of which I chalked up to being just an unpleasant but unique experience that life in Vietnam provided.

When I wasn't assigned to work in the warehouse or hooch areas, I was assigned to other tasks, such as dung and trash burning, guard duty, routine perimeter patrols, and more mundane tasks like kitchen duty. After the first couple of months in Vietnam, my hopes for an education were dwindling fast, but I could peel potatoes with the best of them and learned to burn a lot of—well, to be polite—dung. The fact was, one's assigned military occupational specialty didn't mean much in Vietnam. You pretty much did whatever you were ordered to do—period. No questions asked.

After several months, the high temperatures and mugginess, along with the drudgery of mundane details and warehouse duties, started to wear me down. With no place to go but up, I requested permission to apply for a military driving license, which I was granted after a short proficiency driving test. Once approved to operate military vehicles, I started to work regularly in our company's motor pool, driving 2.5- and 5-ton supply trucks.

Driving afforded me the opportunity to get away from Cam Ranh every now and again. Actually, all things considered, if it wasn't for the war interfering all the time, the countryside was quite picturesque. Now, don't get me wrong; supply runs were still hard work and nerve-racking because of the occasional land mine and sporadic sniper you had to worry about, but it was a change of scenery—a chance to do something different.

All the same, it didn't matter what you were doing; most of the days were still long, hot, and humid. Come to think of it, the nights were long, hot, and humid too. Time went by at what seemed to be a snail's pace. I guess it was the combination of our extended hours and the accelerated workload created by the ever-expanding war and the continually increasing need for critical supplies that made time seem to move so slowly.

During my tours in Vietnam, I participated in a number of convoys, but there's one in particular that stands out in my mind. A group of us were detailed to drive a dozen or so brand-new deuce-and-a-half trucks to Nha Trang, which was just a little over forty-eight kilometers (thirty miles) north of Cam Ranh. We were instructed to leave the vehicles at the depot in Nha Trang, and we would be flown back to Cam Ranh. When I heard that we were going to fly back, I naturally thought that it would be by helicopter. I started to get a little excited about the prospect of flying in a helicopter. Boy, was I wrong; that wasn't going to be the case—at least not that time.

We dropped the trucks off at the depot as instructed and were driven to an airstrip where a Caribou aircraft was waiting for us. The Caribou looked to me to be an amenable little two-prop airplane—a smaller version of the C-130 we'd flown into Vietnam.

We were quickly loaded onto the Caribou by the loadmaster, who had a cantankerous look about him but turned out to be an OK guy. While we were waiting to take off, he told us that the Caribou was the real workhorse of Vietnam; it operated at best under poor conditions, which included the heat, humidity, dust, and mud. All in all, considering we were in the middle of a war zone, the Caribou had an excellent safety record, too—much better than any helicopter, or so the sergeant told us.

Caribou

The Caribou was a great little all-around plane, at least from the information we were able to glean from the plane's very proud loadmaster. But while he told us some of the facts about the little Caribou (for example that it was used to resupply combat units because of its unique ability to fly in and out of difficult areas on short, unimproved airstrips), he neglected to give us one significant piece of information. It seemed that, for combat safety reasons and because of the unique qualities of the short-range Caribou, as the plane neared its destination and was preparing to land, the pilot would turn the engines down to glide the aircraft into its flight path for the landing strip.

Well, there we were, sitting in our web seats, listening to the relentless drone of the twin engines, and then the humming and vibrations suddenly stopped. No warning. No heads-up. Just nothing, but we were still up in the air (hopefully flying).

After the engines were turned down, all I could hear was the rushing whisper of the wind outside. The plane immediately went into a steep dive and onto its approach glide path. It eventually ended up cruising effortlessly onto the landing strip in nearly complete silence until it came in contact with the pierced-steel plank (PSP) runway.

It was a sweet sound when we glided onto the PSP—music to my ears even. As we landed effortlessly, I felt a little bounce and heard the wheels smoothly contact the PSP, signaling that we had indeed landed back at Cam Ranh, and we were still in one piece. In any event, it was better than kitchen or dung-burning duties.

It was another unique experience, another day in paradise, and one more lesson that Vietnam had to impart. It was my first time flying in a Caribou, and I found it very stimulating. While a little unsettling, the short ride in the tiny Caribou turned out to be a thousand times better than flying in a helicopter.

Not long after that, I had my first experience with a Huey. The biggest problem about flying in a helicopter is that they are all about physics and inertial forces like sharp banking turns and bouncing up and down. The only two things standing between you and crashing are the skill of your pilot and one very fragile-looking rotor assembly. If anything were to happen to either of them, it was game over, lights out.

Perimeter Patrol Duty and Unfriendly Contacts

View on perimeter patrol, Cam Ranh 1966

28

Most of the time I was in locations considered by many to be safe and secure, the reality was that no area in Vietnam was genuinely safe or invulnerable. So even though I was not assigned to what I would consider a traditional infantry combat role, Cam Ranh was still frequently attacked by the Viet Cong.

From a military standpoint, the VC considered Cam Ranh and many of the other large US bases primary targets precisely because they were so heavily fortified. They continually bombarded us with rocket and mortar rounds. In addition to their terrorist harassing tactics, there were also frequent full-scale, all-out insurgent attacks, especially during the weeks leading up to Tet Offensive periods.[4]

On many nights, the dark sky was illuminated by bright red or white star cluster flares, and we were put on alert. Although I did feel apprehension about the incoming rockets and mortars, more nerve racking were the swishing and whooshing sounds of our outgoing artillery rounds passing over our location—explosive ordnance intended for some distant inland enemy target(s).

While incoming and outgoing munitions did have different sounds, it's hard to describe the sounds each type of ordnance made. It is something that you experience rather than try to explain, but I'll give it a try. Incoming rockets and mortars didn't make much of a sound before they hit unless they were close or passed over your position. When they did, it sounded like a gigantic piece of paper being torn in two. The pitch of the ripping paper was higher at first and lower as the round passed by or hit the ground. Our outgoing artillery made a sustained, high-pitched streaking noise, similar to, but not quite the same as, the one a jet plane made flying overhead.

Fortunately, like all the various noises of Vietnam (such as explosions, the echoes of our trucks and heavy equipment driving over the PSP, the sound of jets taking off from the Cam Ranh airstrip, and the distinctive, blade-slapping *whop, whop, whop* of the helicopter traffic), eventually all the background noises blended together in a symphony of intermingled sounds. Rackets that became so commonplace we accepted them as a part of our surroundings, and we barely perceived the complex sonata of scrambled clamor.

When we went off base on supply convoys north to Nha Trang or south to Phan Rang, we had to contend with some added problems, (e.g., snipers and mines), but all in all, I had little contact with the VC.

Nevertheless, it was drilled into us to always be on the lookout for them. So while keeping an eye out for "Charlie" was always priority one, it was my encounters with some of the even-less-friendly inhabitants of the country that, for me, were more problematic and dangerous.

The Cold Blooded Adversary

Vietnam is home base for a collection of some of the most hostile insects and vermin on this planet, not to mention some of the most toxic snakes in the world—deadly serpents such as king cobras, along with a horde of diverse vipers, (e.g., the bamboo, Russell's, Malayan, and green pit), and let's not forget about the banded krait or the extremely poisonous family of sea snakes.

The country is also the abode of assorted huge spiders and centipedes. In fact, we'd been told that the venom from the bite of the Vietnamese centipede, though not deadly, would be extremely painful, so you wanted to give those ugly half-foot-long creatures a wide berth. In addition to the nasty, very large predatory arthropods, there were plenty of lizards, rats, bats, blood-sucking leeches, flies, bedbugs, body lice, big cockroaches, water bugs, painful fire ants, malarial mosquitoes, and what seemed like millions of critters trying to take a bite out of you.

Unfortunately, that was one of the primary rationales the military used for having to spray such large quantities of different insecticides in Vietnam—bug sprays ranging from malathion and DDT to chlordane and lindane and way beyond to dichlorvos, diazinon, dieldrin, naled, and Dursban. There was an insecticide to kill just about anything, from bedbugs to body lice. Snakes, on the other hand, were a very different story. Bug spray didn't work on them. Actually, if you used bug spray on a king cobra, it would probably just make the snake nastier, if that were possible.

On more than a few occasions, some less-than-friendly king cobras decided that they liked our cantonment area and tried to move in, but we dispatched them in short order. In fact, I hit a big twelve-footer with a deuce and a half about two klicks out from the warehouse area. The cobra just reared up, hood flared, and struck right at the truck. The deuce and a half was OK, but the cobra was a mess—it lost its head in the truck grill, while its body was still twitching on the ground.

We also had to contend with what many have referred to as "two-step snakes" (green pit vipers) on perimeter patrol details. The little guys liked to slink through the dense foliage above the jungle floor, where you could sometimes catch a glint of their creeping motion as you were keeping a lookout for signs of Charlie in the area. I occasionally saw them slithering around on the branches to the side or in front of me. Seeing the vipers skulking around wasn't too bad, but every now and then, one of the small, bright-green, diamond-headed, big-eyed creatures would fall out of its protective jungle cover, smack dab into your path or onto you, but it would quickly retreat, creeping back into the protection of the dense jungle. As far as I knew, no one in our unit or area was ever bitten by one.

Each and every one of those snakes, insects, and other critters who call South Vietnam their home thrived very nicely in the tropical jungle climate and conditions. Flourishing equally well in that environment were numerous illnesses and ailments, such as malaria, jungle rot and other fungi (e.g., aflatoxin), typhus, dysentery, sunburn, heatstroke, tuberculosis, leprosy, and parasites such as inflammatory liver flukes, which no one has quite gotten a handle on yet.

As the army depot grew and consolidated, so did my rank and tasks. After my trucking days came to an end, I was assigned for a short time to Operation Red Ball requisitions, working as a material release expediter.[5] In due course, I started to settle in and was assigned to the one-person critical supply operation of our company motor pool. From then on, there were no more details or bouncing around; I worked in supply for the remainder of my time at Cam Ranh. Shortly after being promoted to buck sergeant (E-5), I was put in charge of the newly formed short supply resource unit of the Consolidated Cam Ranh Depot. Our group grew steadily from a small two-person setup into a twelve-man, twenty-four-hour-a-day operational unit in what seemed like no time at all.

We would eventually handle all the various critical supplies for the vast Consolidated Cam Ranh Army Depot. We also serviced and filled supply requisitions from the Republic of South Korea Marines—we referred to them as ROKs—and countless expedited Operation Red Ball maintenance, repair, and operations (MRO) requests. We were kept plenty busy by the continually expanding war, but every once in a while, we did get some free time—we called it R and R (rest and recreation)—to take a break and unwind from the grueling pace of our around-the-clock tasks.

After my first tour, I'd earned a thirty-day leave, which I intended to put to good use. My childhood girlfriend and high school sweetheart, Georgia, and I had developed a deep, loving relationship in the short time we'd been dating—a caring bond so strong that not even the Vietnam War was able to break it. Georgia had written to me almost every day while I was in the service. In turn, I wrote back to her as often as I could. Even though we were some nine thousand miles apart, our loving bond grew stronger every day.

Before I joined the army, we'd seriously talked about getting married, but our parents were against it. They kept telling us we should wait because we were too young. So we waited. Georgia went to college in New York, and for my education, I went to the University of South Vietnam. As it turned out, even though we were separated by half a world, toward the end of my first tour of duty, we decided to get married, and more importantly, we were old enough to do it. The war, as always, was keeping me busy, so while I was getting everything together for my flight back to the States, Georgia and her mother had to rush around to get everything ready for our hastily arranged wedding. All I had to do was to get back home in one piece and on time.

My flight home was in a delightful, spacious, modern commercial jet airliner and was uneventful. It was nothing like going to Vietnam had been, and it was a whole lot faster. I had a short layover for paperwork and to get cleared from Fort Lewis, Washington—and I was off. In short order, I was on the last leg of my first journey home. Finally, there I was: home. Georgia and I were married. We went on our quickly arranged honeymoon to Strickland's Mountain Inn in the Poconos, but the honeymoon was over all too soon, and I was on my way back to Vietnam for my second tour.

A few months into my second tour, Georgia and I decided to meet in Honolulu, Hawaii, for my seven-day R and R. Everything pretty much went as planned, except for one small, unexpected but very joyful outcome. We both arrived on schedule and had a great time. While getting a break from the war was excellent, being with my wife for those few days was, of course, the best part of the trip. A couple of months after I returned from R and R, my wife wrote to let me know we were expecting our first child. As best we could calculate, the baby was going to be born somewhere around the end of December 1968. I tried to get the timing right so that after my second tour, I could take another stateside leave when my wife was due. As the nine-month mark approached and my second tour was ending, I again requested and was granted a thirty-day leave to fly home for the birth of our first daughter, Tara.

Everything went well, but as it turned out, we had just one problem. My leave was quickly coming to an end, and Georgia hadn't given birth. I had taken my leave a little too early. As a result, I had to request a compassionate leave extension. After a lot of praying and some arm-twisting and yelling, the army eventually granted me an extra fifteen days of leave, which I'd accrued. Our daughter was finally born on January 16, 1969, just one day before my extended leave was over and I had to start back to Vietnam.

The day after Tara was born, I went to the hospital early in the morning to say goodbye to Georgia and Tara before I left. At that time, I could watch our daughter in the nursery enclosure, but it was against hospital policy for me to hold her. Notwithstanding the foregoing rule, the floor nurses had a different plan, which caught us by surprise.

They dressed me in a surgical gown and mask and let me hold Tara in the nursery for a short time. It was one of the most exceptional experiences and acts of kindness in my life up to that point. After enjoying my time with Tara, I said my tearful goodbyes and started the long, now-routine trip back to Vietnam. Cam Ranh, by this time, had become my home away from home, and this would be my last trip back.

33

Leaving my new family to get on that plane and return to Vietnam was, without question, one of the hardest tasks I've ever had to do in my entire life. There are no words that can even come close to communicating what I was feeling and experiencing during that last flight back to Vietnam. There are no verses I can use to describe it or to put into simple words what I was feeling. Even the idiom "a living hell" doesn't fill the bill.

Once I was back in Vietnam, the time dragged and for the most part is still a shadowy daze. All I seem to recall is that my resource unit had changed locations during my leave, and we were now in more substantial, better-built, two-story barracks similar to stateside barracks but with some critical differences: we still had no indoor plumbing or air-conditioning. So even though the structures changed, the conditions never did.

Our new two-story barracks

Before my trip stateside for Tara's birth, I had been very seriously considering making the army a career. I'd even discussed the possibility with the base reenlistment officer. The re-up officer, after a short review of my personnel file, told me that if I reenlisted for six more years, he

34

could guarantee me, in writing, three years of duty at any army base in the world—my choice—and a $10,000 reenlistment bonus to boot.

Now, in 1969, $10,000 was a lot to consider. In fact, in 2018 dollars, that cash bonus would equate to nearly $70,000, and as an added benefit, if I reenlisted while in Vietnam, it would be income-tax-free. Needless to say, it was quite an incentive and a very tempting offer. But as hard as it was to leave Georgia in 1967, saying goodbye to my new family and returning to Vietnam for that last tour of duty just overwhelmed me.

I realized that while it was a very tempting proposal and a lot of money, having to leave Georgia and Tara again for some stupid war down the road would be next to impossible for me. There was no way I could— or would want to—go through that experience again. So with the birth of Tara and that last tour of duty in Vietnam, any thoughts that I'd entertained about making a career out of the service ended, and the course of my life was altered again.

A few days before I left Vietnam for the last time, I got a little nostalgic and took a long look around the base. Cam Ranh had grown tremendously. It now even had a massive, well-stocked post exchange (PX), replacing the small shack we had back in 1966. Our new PX even had some of the more sophisticated amenities of civilian life, such as pizza— not very good pizza, but pizza nonetheless. The jungle areas of that first sandy plateau we called home were gone; they'd been replaced by two-story barracks, wooden walkways, and PSP. The sounds of the tropical birds, monkeys, and other critters were all gone. Even the snakes had vanished. Almost everything had been transformed from those first formative months I had spent in South Vietnam. What had taken its place was a good-size noisy, bustling, industrialized city.

As hectic as the war was, the last six months still dragged out more than usual. Every day seemed like a week and every week like a month until I finally received my flight orders to the States for separation. As I started looking forward to being discharged and rejoining Georgia and Tara, time seemed to speed up. It began to move a little more quickly as my anticipation and eagerness returned. But although I was excited about returning to civilian life, I would inevitably miss the army and the twelve-man crew I was leaving behind in Vietnam.

As for the education I'd hoped to receive by joining the army, it turned out to be very different from what I'd envisioned. Regrettably, it wasn't until that humid summerlike night in May 2012 that I had begun to fully realize exactly how much Vietnam and the war had affected my life and education. Although it's easy to wax philosophical here and now, almost fifty years later, that was not the case while I was in Vietnam and for years afterward.

Once the dam in my mind broke in May 2012, and I started to recall many of my past experiences, spontaneously replaying them over and over again, I began to understand that I did get an education in the lecture halls of the army and received many lessons in life and responsibility. Essentially, enlisting in the military forced me to grow up fast. The army taught me honor, accountability, and discipline. I was taught patience and perseverance, focus, and commitment in the classrooms of Vietnam. I learned about the best in people and the worst in people. I saw the kindness of man and man's cruelty to man. General Sherman was right when he said:

> I am tired and sick of war. Its glory is all moonshine. It is
> only those who have neither fired a shot nor heard the
> shrieks and groans of the wounded who cry aloud for
> blood, for vengeance, for desolation. War is hell.
> —*William Tecumseh Sherman*

During the almost three years I spent in Vietnam, I was educated in so very many diametrically opposed segments of life and living that perhaps the best way to sum up my experiences and education in-service is with the beginning words of Charles Dickens's novel, *A Tale of Two Cities* (1859):

> It was the best of times, it was the worst of times, it was
> the age of wisdom, it was the age of foolishness, it was the
> epoch of belief, it was the epoch of incredulity, it was the
> season of light, it was the season of darkness, it was the
> spring of hope, it was the winter of despair.

The lessons of life and faith I learned in the military did form a solid foundation on which to build my life. They would inevitably lay the groundwork for my faith in God and the footings on which the rest of my life would be constructed. While I gave a lot of myself throughout those four years, I also received a lot in return. However, it took me a long time to recognize and to appreciate the training I had received.

While most of my tutoring was intangible and on the wild side, I also received an education that wasn't quite so brutal. After leaving the army, with the help of the GI Bill and while working as a police officer, I would have the opportunity to attend college. Eventually, in December of 1979, I received an associate's degree from Fairleigh Dickenson University. So in a roundabout way, I did ultimately achieve my second purpose for joining the army—just not in the way I'd anticipated.

The Last Flight Home

From Vietnam, I flew nonstop to Fort Lewis, Washington, on a rather nice, comfortable commercial jet airliner replete with flight attendants and comfy seats. The trip, which had taken four days in 1966, was now made in a little over half a day—and in complete comfort. The one thing that stands out in my mind about that last flight home was the deafening silence. Everyone on that plane was sitting in their seats somber and, for the most part, expressionless. It seemed like everyone was in a daze—almost zombielike—until we landed in Washington, then the plane erupted, and it was all applause and cheers. We were home.

Once we arrived at Fort Lewis, it took me a few days to complete what I can only describe as a mountain of paperwork and a few medical exams for my release from active duty. Afterward, I received a voucher for the obligatory steak dinner, which, as I recall, was a pretty good meal. But it wasn't my primary interest at the time—getting home was. Finally, on June 22, 1969, with everything completed, I was processed out of the army and taken to the airport for the last segment of my journey home. Everything went well until the jet encountered a series of severe thunderstorms over the Midwest.

Needless to say, it was a bumpy ride, with some pretty significant lightning and downdraft turbulence. All the planes and helicopters that I'd flown in over the past couple of years paled in comparison to that flight. The aircraft lost altitude a couple of times, and it felt like I was back on that Caribou again and coming in for a landing. Only this time, the loss of altitude was much faster and considerably worse. Luckily, we were seat-belted down while everything not secured was flying around the cabin.

Ominously, as you might imagine, the only thing going through my mind over and over again was that after surviving almost three years in Vietnam, I was going to die in a plane crash on my way home. "Now wouldn't that be ironic," I thought to myself. I prayed a lot on that flight. Thank God, we eventually flew out of the storm and landed safely.

Back Home—1969

After I was home for a few days and starting to settle back into civilian life, trying to put the war behind me, my father approached me about filing for disability with the DVA. We had previously discussed at great length my medical issues while in the army, and he knew about the damage done to my lungs, hearing, and ulnar nerve. He felt that all my problems, while not major difficulties at the time, could down the road develop into serious health concerns: smart man, my father.

Unfortunately, I was pretty apathetic about filing for compensation with the DVA—at the time, anyway. Whether my reluctance was because of pesticide exposure, PTSD, survivor's guilt, or just plain old stupidity, I can't really say. What I can tell you is that after several discussions with Dad, a World War II veteran, I was finally convinced that I should submit my medical claims now instead of waiting. I figured that if I filed them immediately, I would always have a permanent record of my assertions on file with the DVA.

I gathered up and filled out all the proper forms along with the supporting paperwork and submitted them to the DVA. As expected, I received a notice a few months later. It was an order to appear for a medical exam. But while I fully intended to follow up on my claims, life happened and got in the way.

I'd just started my new civilian life with my new family in our first apartment with a new job. I just had no time for the DVA and its procedures, especially weekdays between 9:00 a.m. and 5:00 p.m., so I put my service-connected disabilities on hold for the time being. I reckoned that once I was settled into my newfangled civilian life, I'd follow up on those few medical issues. Little did I suspect all the health problems that were waiting to pop up over the next couple of years or that it would be over forty years before I would refile.

Green Pit Viper
Although the Banded Krait was the actual "Two Step" of Vietnam legend, this little guy was no fun to meet up with either.

End Notes - Chapter 3

1. *Operational Report: Lessons Learned*, HQ, Cam Ranh Bay Depot (Prov.) & 504th Field Depot, Vietnam, March 3, 1967.

2. Lineage and Honors, 1st Logistical Command.

3. Ibid.

4. Tet offensives were a series of coordinated attacks on major military installations (Cam Ranh being one those bases) by insurgent forces during the war. These North Vietnamese attacks were timed to coincide with their Lunar New Year, which they called "Tet."

5. Operation Red Ball was a special supply and transportation system established by the Office of the Assistant Secretary of Defense on 1 December 1965. The program was designed to be used in lieu of normal supply procedures exclusively to expedite repair parts needed to remove critical equipment from deadline status. Red Ball Express requisitions were prepared by the needing unit and sent to the nearest Direct Support Unit for supply. If the Direct Support Unit such as the 423rd could not fill the request, the requisition was forwarded to the supporting depot. If the depot could not satisfy the request, it was forwarded to the Red Ball Control Office in Saigon, where in-country assets would be searched for available stock.

Chapter 4
Medical Problems Begin

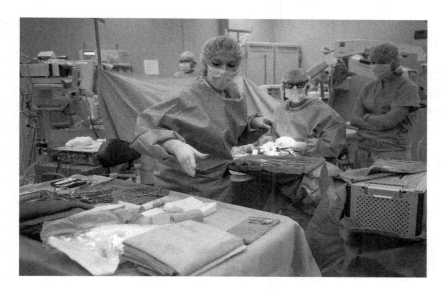

My first post-service medical problem emerged rather modestly, shortly after I left the army, in the fall and winter of 1969. I developed an annoying rash on my face, and as a matter of course, I attributed it to the dry, cold weather having a detrimental effect on my skin. It was, after all, winter, following almost three years in a tropical jungle environment; I was pretty sure I wasn't acclimated to the cold northeastern US climate yet. I treated the rash with moisturizers and skin conditioners, but they didn't help.

Second, I'd been experiencing what I believed at the time to be minor indigestion/heartburn, with a few other minor digestive problems for which I took Alka-Seltzer and antacids. They both seemed to help with what I assumed to be every day, commonplace digestive issues.

Third, I'd developed another bad upper-respiratory infection and had to go to our new family physician, Dr. Crystal. He asked me some questions, looked me over, and told me I had allergies (what I would term hay fever), another upper-respiratory infection, and chronic bronchitis to boot.

In 1969–1970, all my medical problems were annoying and sometimes painful, but they were treatable and not life threatening. Unfortunately, I didn't have medical insurance back in the early 1970s, and money was tight. All the same, as time went on, little by little, I started to have more intense digestive problems, such as some severe bouts of indigestion, in addition to sharp pains in my chest, which I attributed to gas at the time.

My chest pains were intense although intermittent, and taking Alka-Seltzer and antacids usually took care of the discomfort. The rash on my face continued to be red, itchy, and scaly—not a significant problem but annoying. Eventually, I found time and money to see a dermatologist. It turned out the rash was seborrheic dermatitis (a.k.a. seborrheic eczema). So I naturally asked my doctor, "What is it, and how did I get it?" He told me that he didn't know, but it was suspected by researchers that a yeast found on everyone's skin called *Malassezia* combined with facial oil secretions and the immune system all had a hand in causing the scaly inflammatory rash on my face.

After a little reflection on what the doctor said, my mind jumped from the itchy skin rash to Agent Orange and Vietnam. However, our government was continuing to deny that any medical problems or skin conditions—other than chloracne—resulted from our exposure to Agent Orange. In the end, there was just no contemporary evidence that the herbicide was connected to my particular type of dermatitis; nevertheless, my mind went there anyway.

My upper digestive problems gradually became worse, with severe pains in the center of my chest. In fact once, around 1973–1974, I thought I was having a heart attack and was taken to the emergency room at Holy Name Hospital in Teaneck, New Jersey. The emergency room (ER) doctor told me my heart was fine and the problem was probably indigestion. He suggested that I follow up with a gastrointestinal (GI) specialist. Following the ER doctor's advice, I went to see a GI expert.

After a short examination, he diagnosed me with achalasia. While I didn't want to look stupid, I had to ask, "So what is achalasia, and how did I get it?" The doctor told me it was a rare disorder (incidence of one in one hundred thousand) that made it difficult for food and even liquids to pass into my stomach because the nerves at the bottom of my esophagus had become damaged and unable to function. Basically, my esophagus had lost its ability to squeeze food down into my stomach. The muscular valve between my esophagus and stomach, called the lower esophageal sphincter, or LES, didn't fully relax, making it difficult for food to pass into my stomach.

"So what caused the achalasia?" I asked inquisitively. The doctor told me no one was sure what caused the disorder. What he did know was that for some still-unidentified reason, the nerves that opened and closed the LES muscle gradually deteriorate and died, which meant that the LES would fail to open correctly.

The doctor went on to say that according to medical research, the cause could have been inherited, autoimmune, or environmental; it could also have a viral component or could even be a combination of all four factors. Here again, after a short reflection, the one thing he didn't mention—Agent Orange—sprang back to mind, again for the same reason as before. Now my medical problems were beginning to seem a little less commonplace and a lot more worrisome. The doctor finished by telling me that my achalasia would not get better but that it probably shouldn't get any worse either, given the length of time I'd had symptoms.

As I quickly found out, the major problem with achalasia was getting food into my stomach. In time, I discovered that the most effective way to do that was just to stand up when I ate and let gravity work. While standing up did the trick most of the time, once in a while, I had to jump up and down to get food into my stomach. In addition, I would frequently have to take medication to soothe the heartburn-like irritation I would experience.

Unfortunately, the LES continued to deteriorate. Finally, it closed tightly and stayed that way. The end result was that no food was moving into my stomach no matter what I did; it was all staying in my esophagus. At that point, I was left with only two options, and both choices were terrible.

My first choice was an esophageal balloon-dilatation procedure, which is a surgical technique that is exactly what it sounds like. The doctor proposed to stick a deflated balloon down my throat and blow it up when it reached my LES, all in an effort to hopefully tear and stretch my LES open. The second possibility was even worse—a Heller myotomy (before laparoscopic surgery)—where the doctor would have to cut me open, remove a few ribs, and surgically cut or, as he put it, "cleave the LES muscle."

In April 1979, Dr. Newton Scherl performed a successful lower esophageal balloon dilatation. So with my LES ripped, allowing food to get into my stomach more easily, my bout with gastroesophageal reflux disease (GERD) was just beginning. At least I could eat comfortably without standing or jumping up and down all the time, but the heartburn was horrendous. In fact, a little while after I had my first meal following the balloon procedure, I felt like I'd swallowed a hot coal. The pain was that intense, even with my hospital-strength antacids.

Before my achalasia surgery, I almost never went for annual checkups or took proper care of my health, mainly because of costs. We always put our four children first, so the only time I would see a doctor was when there was a medical problem I couldn't solve. However, after the esophageal dilatation, I started to go for regular checkups, seeing my primary care physician (PCP) a minimum of once a year.

In the early to mid-1980s, thanks in large part to my regular checkups, I was told by my PCP that he was concerned about my rising blood pressure readings. This concern was quickly followed by unhealthy rises in cholesterol, triglycerides, and blood sugar. So, using a little common sense, I became more active; joined King Arthur's Gym in Fort Lee, New Jersey; and started to exercise on a regular basis. In addition to aerobics and weight training, I also kept a close eye on my diet. While trying to stay fit and watching what I ate helped with some issues, I continued to have respiratory problems, allergies, and GERD.

In February 1991, I developed a severe sharp pain in my right side and thought it might be appendicitis, so I went to be checked out by my PCP, Dr. DeLuca. He told me that my appendix wasn't the problem and that he could find nothing wrong with me. The pain subsided for a while but would kick up on occasion. It would come and go and could only be described as a pain you might associate with appendicitis.

By November 1991, in addition to the sporadic pains, I started to have dark red blood in my stools. I went to Dr. DeLuca again, who in turn referred me to my gastrointestinal specialist. So it was back to Dr. Scherl.

Dr. Scherl diagnosed me with ulcerative colitis (UC), for which he prescribed a couple of different medications. While I was thankful it wasn't cancer, his finding was still disconcerting. He anticipated my first question and began by telling me that UC was an inflammatory bowel illness that affected the innermost lining of my large intestine, causing long-lasting inflammation and ulcers (sores) to form in my intestines. The doctor continued on to say that UC doesn't develop suddenly but advances rather slowly, over an extended period.

Once again, my curiosity got the better of me, and I had to ask, "What causes UC?" Dr. Scherl replied with the same confounding statement I'd heard from many of my other doctors: he didn't know the exact cause or causes. He continued on to clarify his statement by saying that although no one was sure of the reasons the immune system malfunctions and the intestines become inflamed, there was a firm belief that UC was caused by a combination of several factors, such as stress, environmental contaminants, and genetic predisposition. He asked if anyone in my family had UC. I told him, "Not as far as I know." Even my father and mother didn't recollect anyone in our family having UC, or achalasia, for that matter.

The colitis stayed reasonably stable on the medications but would flare up every so often. Once again, I wondered if Agent Orange could be involved. Regrettably, it wouldn't be until I started my research in 2012 that I would actually discover that Agent Orange was just one of several military pesticides that were—to use the DVA's terminology—"more likely than not" very capable of negatively affecting my immune system, as well as every other biological system in my body.

Then rather suddenly in July 1993, the UC became exacerbated. My intestines became severely inflamed and extremely painful. This time, the attack was bad enough for Dr. Scherl to admit me to Englewood Hospital. But the UC continued to worsen, even with the hospital's steroid intravenous treatments.

The illness devastated my large intestine—so much for slow moving. By this time, I was in some deep, serious trouble. The UC was just eating up my bowels, and I was losing a lot of blood. Dr. Scherl had no choice but to transfer me to the GI experts at Mount Sinai Hospital in New York. I was there for only a day or two, while they ran a lot of tests and took a lot of pictures of my innards. Eventually, I was diagnosed with severe fulminate ulcerative pancolitis and toxic megacolon with a perforation of my large intestine. With peritonitis setting in, Dr. Joel Bauer performed what he termed an emergency ileostomy.

Between 1994 and 1996, after the removal of most of my large intestine, I had four additional abdominal surgeries to try to correct various problems that had occurred. After a long recovery from my last surgery, which was in late 1996, I began to develop what I thought was a pimple on my lower right leg in January of 1997. The pimple started to spread and transformed into an open sore for no apparent reason.

I initially believed that I'd scratched it and that it had become infected—not just a simple, ordinary infection but a nasty, fast-moving, flesh-eating infection like methicillin-resistant *Staphylococcus aureus* (MRSA), the infection that killed Muppet creator Jim Henson in 1990. In no time at all, I had the same problem, only this time on my lower left leg. Yep, it was time to see my doctor and hope that it wasn't too late.

After having to wait for two agonizing days, I finally got to see Dr. Gary Bauer and was diagnosed with pyoderma gangrenosum (PG). Well, at least it wasn't MRSA, although it sure acted like it. Once again, I just had to ask, "So what is pyoderma gangrenosum, and what did I have to do to get it?"

Dr. Bauer told me that it was a rare condition—once again, occurring in about one out of one hundred thousand people—that caused large sores or ulcers to develop on the skin, usually on the lower legs. Unlike UC, PG is a very fast-spreading and destructive inflammatory disease. The spreading wound produced by PG can occur with or without any apparent underlying disorder or association, but in my case, there was a primary cause. The doctor went on to tell me that some of the diseases associated with PG were UC, polyarthritis (inflammation of several joints), gammopathy, vasculitis, leukemia, and several other conditions. Of course, my mind zeroed right in on leukemia and Agent Orange. While I was still contemplating that thought, he went on to explain that PG is actually in the family of autoinflammatory skin diseases called neutrophilic dermatoses.

The doctor must have seen the blank look on my face that indicated I didn't know what the heck he was talking about. This time, before I could even open my mouth, he was already explaining with a little smirk how a neutrophil is a type of white blood cell or leukocyte that forms as a line of defense against bacterial infections. As a consequence, my extended treatment would involve wound care as well as the use of anti-inflammatory agents, including antibiotics, corticosteroids, and immune-suppressant drugs.

Dr. Bauer advised me that my PG ulcers were not from leukemia but from an immune system reaction attributed to my history of UC and the presence of the tiny portion of my colon that remained. After several months of intensive treatment, the ulcers on both my legs were healed.

Nevertheless, Dr. Bauer recommended that I have the remainder of my colon removed, or the PG might reactivate, and the ulcers return.

The operation to remove the rest of my colon was done by Dr. Ronald White at Englewood Hospital in late 1997. After that last surgery in 1997, and with all the diverse medical issues that continued to affect my health, I finally decided to retire earlier than I had anticipated. Thus, in December of 1999, I retired from police work somewhat reluctantly. From my retirement until I reached Medicare age, my health stayed pretty stable with no significant new problems—a fact I wholly attribute to retiring early. Had I continued to work as a police officer, I'm not sure I could have said the same thing.

Anyway, when I turned sixty-five, I went for my "welcome to Medicare" physical exam. Everything went as expected and seemed quite routine. However, shortly after returning home from the exam, I received a frantic telephone call from my PCP's office. The nurse told me that Dr. Chon needed to see me right away, today. When I asked her why, the nurse could only tell me, "Today, now!" So I headed back to the doctor's office, thinking all the way about what the heck could possibly be wrong.

Dr. Chon told me that during the electrocardiogram (EKG), they discovered something called atrial flutter (AFL), which is an abnormal heart rhythm. It is similar to atrial fibrillation (AFIB), but with AFL, the irregular heartbeats are very rhythmic. AFL is a rapid beat in the atria of the heart or, as my doctor termed it, a supraventricular tachycardia. According to Dr. Chon, the upper chambers of my heart were beating way too fast, which resulted in atrial muscle contractions that were out of sync with the lower chambers of my heart, or my ventricles. He went on to tell me there was no way for him to know how long my heart had been out of sync but that I was fortunate that it had been discovered. In fact, he said that because it was my lucky day, I should play the lottery. According to the doctor, because my heart was beating normally at 70 beats per minute (BPM), but my atria were flapping at over 180 BPM, there was no way he could have or would have found the AFL without the Medicare EKG. He told me I was blessed that I didn't have a stroke or, as he put it, a cerebrovascular accident, and he put me on blood thinners.

I asked, "So now, what did I do to get this type of heart problem so suddenly?" hoping to make light of something my doctor apparently didn't think was all too funny. With his usual concerned expression, he told me that in all probability it was my chronic bronchitis (COPD) that had triggered my heart to malfunction. He went on to explain to me that I could leave it alone and stay on blood thinners for the rest of my life, or I could have a medical procedure known as cardiac ablation.

So once again, my choices went from bad to worse, but I didn't want to stay on blood thinners for the rest of my life either. So I asked, "What does a cardiac ablation involve?" He told me that it was a specialized surgical procedure used to correct abnormal heart value rhythms such as mine. It worked by scarring or destroying the damaged tissue in my heart that was triggering the abnormal atrial beats. He went on to say that the procedure involved the insertion of a long, flexible tube called a catheter through a major vein or artery in my groin and threading it through until it reached my heart. The surgeon would then burn or freeze the tissues that were causing the abnormal electrical impulses. After doing some research on AFL and experiencing life on blood thinners, I decided to have the ablation. Everything went well with the surgery. The only problem is that now I have a fifty-fifty chance of developing AFIB because of the AFL and subsequent cardiac ablation.

From my retirement in December of 1999 to the present, my COPD, hypertension, endocrine issues, and other medical problems have slowly but steadily progressed, getting a little worse each year. Over time, I have had greater difficulty controlling my growing list of ailments with just diet and exercise. As a result, several medications had to be added to my daily routine just to maintain a semi-healthy life.

Well, that does it for my trip down memory lane. I could go on and regale you with more stories about my time in Vietnam and added accounts of my medical issues, but I think you have a pretty good condensed version of what transpired between 1969 and 2012. Besides, while this book is about me and my recollection of Vietnam, it's also about what I found out during my investigative research into the multitude of toxic chemicals and conditions I and so many others were exposed to in Vietnam and the ensuing clandestine administrative cover-ups.

When I began my research in 2012, much to my annoyance and surprise, I discovered that not only was my exposure to Agent Orange problematic, but I'd also been exposed to several other equally hazardous and damaging herbicides and insecticides. Toxic pesticides just as biologically harmful and toxicologically devastating as Agent Orange, but less publicized. It's almost as if these other shadowy, unpublicized pesticides were still being guarded from disclosure or, even worse, just being ignored, hoping they would go away. The old "Delay, deny, until they all die" governmental whitewash strategy.

The investigation I started in 2012 would, in the end, lead me to the conclusion that most of my laundry list of ailments—to use the DVA's jargon again—were "more likely than not" caused by my exposure to all the dangerous pesticides used in Vietnam, Agent Orange being only one of several.

However, before I delve into all the data my research uncovered, there are two toxic contamination events I would like to highlight in this work. I learned about both these incidents during the course of my studies and, while they are very thought provoking, they just simply astounded me. Not only did they shock me, but they also made me quite angry. Not because of the events themselves but because of the reactions of our government and military leaders.

Chapter 5
Dioxin-TCDD Incidents

Over the course of my research on Agent Orange et al., I've examined many dioxin and other toxic chemical exposure incidents. While I found many of those events to be troublesome, two were particularly upsetting. The first is an episode that occurred in 1976—shortly after the Vietnam War—called the Seveso Incident.

On July 10, 1976, the chemicals 2,4,5-T (a chemical contained in Agent Orange) and dioxin-2,3,7,8-TCDD (a highly toxic component of Agent Orange, although a manufacturing contaminant) were accidentally or carelessly released when an explosion occurred at an industrial plant near the small town of Seveso in northern Italy.

The chemical plant was owned by Industrie Chimiche Meda Società Azionaria (ICMESA), a subsidiary of Givaudan, which was, in turn, a subsidiary of Hoffmann-La Roche. At the time of the incident, the ICMESA plant was manufacturing the organic chemical 2,4,5-trichloro-phenoxy-acetic acid (2,4,5-T), and for some still-undetermined reason, a rapid, unanticipated rise in temperature occurred in one of the company's chemical reactors.

This increase in temperature caused a surge in pressure, eventually resulting in the failure of a safety valve, followed by a sudden explosion. The blast released a cloud of chemicals containing 2,4,5-T and its contaminant, dioxin-2,3,7,8-TCDD, into the atmosphere. The expanding plume of chemicals eventually contaminated a densely populated area downwind from the discharge site.

The exact mass of the deadly dioxin-2,3,7,8-TCDD released by the detonation remains unknown, but it's estimated that between one and two kilograms were released into the atmosphere—equivalent to one to two billion micrograms (μg). The actual area affected by the extensive contamination, roughly seven miles or twenty-five million square yards, was quickly cordoned off and split into three zones (A, B, and R) based on the decreasing concentrations of dioxin-TCDD found on the soil surface.

Zone A—the most heavily tainted area—was determined to have a dioxin-TCDD mean soil concentration of 230 μg per square meter (μg/m²) and was populated by roughly 736 residents. As a point of reference, a single poppy seed weighs about 0.3 mg or 300 μg. That's 23.3 percent heavier than the mean soil concentration of dioxin-TCDD in Zone A.

Zone B was determined to have a dioxin-TCDD soil concentration ranging between 5 and 50 μg/m² and was populated by roughly 4,700 residents. Zone R had an almost negligible dioxin-TCDD soil concentration of fewer than 5 μg/m² and a population of approximately 31,800 residents.

The inhabitants of the impacted zones were advised to destroy and not consume any locally grown fruits or vegetables, which would be banned from consumption for several months. Within days of the accidental/careless release, around 3,300 animals were found dead, mostly small animals such as poultry and rabbits.

By the end of August 1976, Zone A had been evacuated and fenced off, while at the same time, emergency slaughtering of livestock commenced, preventing dioxin-TCDD from entering the food chain. The available data indicates that over eighty thousand animals were killed and disposed of as a result of that purging and the contamination concerns of the Italian government. Two senior executives of the company, Herwig von Zwehl, and Paolo Paoletti, were considered responsible for the accidental or careless discharge. They were both arrested by the Italian government and ordered to stand trial.

Paoletti was assassinated in 1980 before his trial by a left-wing group called Prima Linea, while Zwehl and four other employees of ICMESA or Givaudan were eventually tried in Italian courts. All five were found guilty and sent to prison in 1983; they received sentences ranging from two and a half to five years in jail.

In addition to the court trials, two commissions were established by the Italian government to develop plans for the quarantine and long-term decontamination of the three affected zones. What's more, the Italian government set aside 40 billion lire, or $47.8 million (in 1976 US dollars), for the remediation and overall cleanup of the contaminated areas. Unfortunately, the cost of the actual decontamination effort skyrocketed, and the initial estimated cost would be tripled just two years later.

All these emergency actions, disruptions, precautions, and future long-term actions and health studies were the responses of the Italian people and government—in 1976—to just *one* unintended release of the chemicals 2,4,5-T and dioxin-TCDD into the atmosphere. Only two harmful chemicals, spread by natural air currents, which managed to contaminate over seven square miles of Italy on one warm day in July 1976, with dioxin-2,3,7,8-TCDD measuring less than the mass of one poppy seed spread over each square meter (1.09 yards) of topsoil.

So what did the Italian government know in 1976 that the US government didn't know about the Seveso Incident and the far-reaching 2,4,5-T and dioxin-TCDD contamination?

Now for the part that made me angry. In July of 1976, here in the United States, while our military, government, and the DVA were celebrating our bicentennial, they were also telling in-country veterans— and forcefully maintaining—that there was "no evidence" or "insufficient evidence" of any illnesses, harmful effects, or adverse health impacts regarding our multiple unprotected pesticide exposures while we were serving in Vietnam. So while the Seveso Incident precautions were being mandated, planned, and executed, and two senior executives responsible for the accidental/careless release were arrested, we were unfortunate casualties of governmental shame and the congressional budget process. I would have to guess that our leaders believed that what happened in Seveso, stayed in Seveso.

Conceive, if you can, what the reactions would have been in Italy—or in any civilized country, for that matter—upon learning that after being deliberately atomized, an herbicide was purposefully spewed into the

53

atmosphere from low-flying aircraft moving at 150 to 170 miles per hour over a period of roughly six years—2,190 days. An herbicide containing literally millions of gallons of not only 2,4,5-T but also the chemical known as 2,4-D (2,4-dichlorophenoxyacetic acid), accompanied by over 366 billion μg (366 kg) of the toxic contaminant dioxin-TCDD.

Visualize just these three harmful chemicals being mixed with jet fuel (JP-4) containing even more exceptionally harmful ingredients—substances such as benzene, toluene, ethyl-benzene, and xylene, paraffin, and polycyclic aromatic hydrocarbons. Now picture that after the pesticide had been combined with the JP-4, the finished product was then loaded into specially designed tanks that would atomize the mixture into a fine mist, realizing all the while that this process would create a toxic chemical fog capable of being carried by natural currents of air over millions of acres of land.

But wait, we don't have to imagine that scenario at all. It has already happened, and the herbicide was called Agent Orange. While that alone is a whole lot to contemplate, I'm not finished yet. What's more bloodcurdling are the responses of our government and military leaders in 1976 to these documented and verified multiple releases of injurious chemicals during the Vietnam War. Their reactions were to deny, deny, deny and then more forcefully to reject the very existence of these toxic pesticides and associated chemicals. They also blocked all attempts to medically link any cancers and critical illnesses with exposure to any of those pesticides. In fact, they would reject all responsibility for any of the illnesses that started to manifest in thousands of veterans after the war and for decades later.

Our government and military leaders left veterans unprotected from their planned deployments and aerial dispersal of all the multicolored, contaminated tactical pesticides used during the war. They made no arrests. No one went to prison over the deliberate release of these health-damaging—and in some cases, internationally banned—hazardous substances. The reactions of US officials were very different from the protective measures characterizing the Seveso Incident. Our bureaucrats and military undertook no emergency slaughtering of animals to prevent dioxin-TCDD or any of the other dioxin-like compounds (DLCs) from entering the food chain. In fact, quite the opposite was true. We veterans were supplied with water, fish, meat, fresh produce, and milk products that were raised, grown, and procured in South Vietnam.

54

No safeguards were taken to prevent dioxin-TCDD or any other toxic chemicals from entering our water supply. No precautionary studies explored the negative health impacts of exposure to the tactical military pesticides before they were deliberately deployed. No protection was provided, and there was no apparent health concern for the military personnel who were exposed to the onslaught of hazardous chemicals.

The US government and the DVA initially rejected and denied any of the serious health problems that began to occur among the tens of thousands of veterans who served in Vietnam, simply because we veterans couldn't prove a medical connection for the cancers, illnesses, and disorders. The government denied them all, even though they knew or should have known the true nature of the systemically damaging pesticides they allowed to be deployed in Vietnam.

Our leaders knew and could corroborate the high toxicity and insidious systemic health-damaging nature of the chemicals veterans were exposed to, but they kept the information shrouded in secrecy. They would use contrived national security concerns as a rationale for that suppression. To compound their deceptive actions, they repeatedly refused to study the health of veterans who had been regularly exposed to their wide-ranging chemical barrages after the war.

For the most part, the military pesticides were stored in metal fifty-five-gallon barrels with color-coded stripes as identifying markers. The government and military provided little factual and no cautionary information about their dangerous pesticides. Yet in spite of official governmental denials, many important letters, reports, and official documents exist that confirm that our national leaders and many chemical company executives from firms such as Dow and Monsanto knew of the systemic health problems that would be produced by the toxic pesticides they allowed to be purchased by the military and unleashed on military personnel during the war.

One such report is the now declassified 1990 document compiled by Admiral Elmo Zumwalt, who confirmed that US military experts knew that the pesticide Agent Orange was systemically damaging at the time it was being recommended for use in Vietnam. Admiral Zumwalt also documented in his report a 1988 communication from Dr. James R. Clary, a former scientist with the Chemical Weapons Branch, to Senator Tom Daschle. The following excerpt is in Dr. Clary's actual spine-tingling words:

When we (military scientists) initiated the herbicide program in the 1960s, we were aware of the potential for damage due to *dioxin contamination* in the herbicide. We were even aware that the "military" formulation had a *higher dioxin concentration* than the "civilian" version due to the lower cost and speed of manufacture.

However, because the material was to be used on the "enemy," none of us were overly concerned. We never considered a scenario in which our own personnel would become contaminated with the herbicide. And, if we had, we would have *expected our own government to give assistance to veterans so contaminated.*[1]

What more needs to be said? I think Dr. Clary's own words very persuasively corroborate the fact that our leaders knew the true nature of the pesticides they allowed to be spewed on us, and they let them be used anyway while holding everyone involved to enforced silence and classified secrecy.

Long-term Seveso Follow-up and Other Studies

The concern for the Seveso residents exposed to 2,4,5-T and dioxin-TCDD in 1976 was so great that there were planned long-term (twenty-and thirty-year) health studies. This follow-up research broadly revealed a higher mortality rate from cardiovascular and respiratory diseases. More specifically, these examinations discovered statistically significant increases in chronic obstructive pulmonary disease, chronic ischemic heart disease, hypertension, and diabetes within the exposed residents when compared to the non-exposed control population. Seveso scientists went on to postulate that not only did the 2,4,5-T and dioxin-TCDD have negative health impacts on the Seveso residents, but the stress of the incident itself had significant deleterious health and other adverse effects.[2]

Ultimately, follow-up investigators concluded that it was the genotoxic nature of the chemicals the Seveso inhabitants were exposed to that would best explain the long-term adverse health consequences and numerous early deaths reported in the long-term studies.[3]

While I have highlighted the Seveso aftereffect studies, many other relevant, mostly foreign studies were undertaken. One such study was conducted with over one hundred thousand Korean in-country Vietnam War veterans between 2000 and 2005. This study offered some startling as well as impressive results. For example, the Korean soldiers who served in Vietnam had a higher incidence of thyroid disorders, type 2 diabetes mellitus, and other endocrine disorders as well as systemic atrophies, including spinal muscular atrophy, peripheral polyneuropathies, angina, stroke, COPD—including chronic bronchitis and bronchiectasis—asthma, peptic ulcer, Alzheimer's disease, and cirrhosis of the liver.[4]

An entirely different but parallel Korean study found a strong association between Agent Orange exposure and disorders of the endocrine, immune, and nervous systems in Korean veterans.[5]

Additionally, one study was even conducted using actual US Army veterans (three decades after their service) who were occupationally exposed to phenoxy (2,4-D) herbicides while serving in Vietnam. The study revealed that these exposed veterans had a statistically significant higher incidence of diabetes, heart disease, hypertension, and nonmalignant lung diseases when compared with veterans who had not been exposed to herbicides and conditions of Vietnam.[6]

I could go on and on with several other astonishing, mostly foreign, studies of veterans exposed to the same toxic chemicals US veterans were, but I think I have made my point with just these few documented studies. But, there is still one more event that I would like to cover before I move on. This incident was not a study but a buyout.

Times Beach Relocation Project

In 1982, right here in the US, there was a dioxin contamination incident—just one of many[7]—at Times Beach, Missouri. It was discovered that the small rural town had been polluted with dioxin-TCDD. After the discovery, the National Institutes of Health confirmed that Times Beach had measured dioxin surface soil concentrations ranging from 100 parts per billion (ppb) to a high of 317 ppb. Subsequently, as a result of those extraordinarily high dioxin levels, on February 22, 1983, a voluntary buyout was announced by the Federal Emergency Management Agency (FEMA).

This buyout, in turn, resulted in the homes of all two thousand or so residents, along with their businesses, being purchased by the US government. In effect, this allowed the whole town to be evacuated and moved, with taxpayers footing the bill.[8]

Before the buyout, the commonly accepted level of dioxin-TCDD contamination was set by the Centers for Disease Control and Prevention (CDC) at 1 ppb. The reason the CDC had calculated this 1 ppb level was that previous lab tests had shown that the LD_{50}—50 percent lethal dosage—of dioxin-TCDD was only 0.6 micrograms per kilogram of body mass in guinea pigs. So at the time, the CDC felt that 1 ppb was an appropriate level of protection for TCDD contamination. Today (2019), however, the allowable contamination level for TCDD is set considerably lower at 0.00000003 milligrams per liter (mg/L).

All dioxins (including TCDD) have been classified as known human carcinogens by many scientific organizations. They have also been acknowledged to have numerous noncancerous health effects as well. Some of these adverse health impacts are atherosclerosis, hypertension, and diabetes. In fact, it has been medically accepted that long-term contact with low levels of TCDD disrupts our nervous, immune, reproductive, and endocrine systems.[9]

So to recap, we have all the previous studies and reports, plus the many others that are not recorded in this work, and last but not least, the Times Beach buyout project. We also know Times Beach had measured TCDD levels ranging from 100 ppb to 317 ppb, much higher than the safe level generally accepted by either the CDC or the Environmental Protection Agency (EPA) for that time frame.

Unluckily, we boots-on-the-ground Vietnam veterans and the residents of Seveso were exposed to dioxin contamination levels considerably higher than those measured at Times Beach. In fact, according to the latest research calculations on just the Agent Orange used in Vietnam, on average, it contained 13 parts per million (ppm) of dioxin-TCDD, which is equivalent to 13,000 ppb, or about forty-one times greater than the highest dioxin levels measured at Times Beach.[10] Even worse, other sources have reported that the dioxin-TCDD levels in many shipments of Agent Orange measured as high as 60 ppm or 60,000 ppb.

Once again, just as a point of reference for your consideration, fifty-two drops of ink in a fifty-five-gallon (208-liter) barrel of water would produce an ink concentration of roughly 13 ppm.

Dr. Stellman and her colleagues have conservatively estimated that the total dioxin-TCDD content in the Agent Orange sprayed within South Vietnam during the war was roughly 366 kg (807 pounds).[11] Unfortunately, these significant statistics do not include TCDD and DLCs from other herbicides (e.g., Agent White), insecticides (e.g., malathion and DDT), or dioxins created from dung and trash burning.

The bottom line is in 1982–1983, at the same time the US government was telling veterans that there was no proof or insufficient evidence of any cancers, illnesses, or other health problems resulting from their exposure to all the pesticides used in Vietnam, they purchased a whole town in response to significantly lower exposure levels to just one toxic organic chemical, dioxin-TCDD. The governmental chant of "insufficient evidence" and the Times Beach buyout all happened in the same year that five people were being sentenced to jail in Italy for the Seveso Incident.

In an effort to be clear, and contrary to how the word is commonly used, dioxin is not one chemical but a very complex, interwoven family of ingredients consisting of various dioxins, furans, and polychlorinated biphenyls (PCBs). All the members of this vast chemical family—at least seventy-five different compounds typically referred to as polychlorinated dioxins—have a wide range of toxicity, with the dioxin we know as 2,3,7,8-TCDD being the most hazardous and lethal of the group. Nonetheless, all the chemicals in the polychlorinated biphenyls family are harmful to humans in varying degrees from bad to worse to downright evil.[12]

Unequivocally, no dioxin is safe or harmless. Dioxin is definitely not found in a pure state, and it's certainly never found alone in our environment. To further exacerbate the impacts of dioxin, when it's formed in our environment, it's spawned along with plenty of unwelcome, unfriendly toxic companions. All of which are very capable of negatively affecting the health of any human exposed to them.

What we term *dioxin* is, in reality, a combined multifaceted and variable mixture of CDDs, furan, and PCB components all wrapped up into a myriad of complex, interwoven, dangerous organic substances that will adversely affect the cellular structure and health of anyone unlucky enough to be exposed to them, even in minuscule doses—amounts smaller than a granule of table sugar.

Now is where I really started to have some fun by putting the 1960s' DuPont tagline "Better things for better living through chemistry" to the test. As you read the next few chapters, you may become a little weary and frustrated with all the scary-sounding, hard-to-pronounce chemical names and acronyms, but remember that in-country veterans were exposed to all those very same hard-to-say, scientific-sounding terms. It may even seem to some of you like a chemistry course gone wild. Unfortunately, as much as I would like to make the chemicals we were exposed to easy and uncomplicated, I can't.

All the chemicals in the following chapters have been documented to have been contained in the tactical pesticides—both herbicides and insecticides—that were aerosolized into a fine mist and unquestionably sprayed within and near Cam Ranh during the war. While these same chemicals were used throughout Vietnam, I can only address the ones I know were deployed in my locations.

As you continue reading, remember that this is a relatively small collection of some of the most prevalent pesticides and their toxic companion ingredients used during the Vietnam War. A full list of all the pesticides used by the military is contained in the *Naval Medical Field Research Laboratory List of Herbicides and Pesticides Used in 1968 by US Armed Forces*.

End Notes - Chapter 5

1. Department of Veterans Affairs: Report to the Secretary of The Department of Veterans Affairs of the Association Between Adverse Health Effects and Exposure to Agent Orange. By Admiral Zumwalt

2. "Dioxin Exposure and Non-malignant Health Effects: A Mortality Study," *Occupational and Environmental Medicine* 55, no. 2 (1998): 126–31. A. C. Pesatori, C. Zocchetti, S. Guercilena, D. Consonni, D. Turrini, and P. A. Bertazzi

3. "The Seveso Studies on Early and Long-Term Effects of Dioxin Exposure: A Review," *Environmental Health Perspectives* 106, Suppl. 2 (1998): 625–33. P. A. Bertazzi, I. Bernucci, G. Brambilla, D. Consonni, and A. C. Pesatori

4. "Agent Orange Exposure and Disease Prevalence in Korean Vietnam Veterans: The Korean Veterans Health Study," *Environmental Research* 133 (2014): 56–65. S. W. Yi, J. S. Hong, H. Ohrr, and J. J. Yi

5. "Association between Agent Orange Exposure and Disease Prevalence of Endocrine, Immune, and Nervous System in Korean Vietnam War Veterans: Korean Vietnam War Veterans Cohort Study," *Organohalogen Compounds* 73 (2011): 1468–71.

6. "Health Status of Army Chemical Corps Vietnam Veterans Who Sprayed Defoliant in Vietnam," *American Journal of Industrial Medicine*, Online First, September 27, doi:10.1002/ajim.20385.

7. Love Canal and the Poisoning of America - The Atlantic Magazine 1979

8. U.S. Department of Health and Human Services, *Report on Carcinogens*, 12th ed., 2011, citations excluded; 317 PPB would be equal to roughly 317 drops of ink put into a large tanker truck, about 11,000 gallons, of water.

9. "Dioxins and Human Toxicity," *Archives of Industrial Hygiene and Toxicology* 61, no. 4 (2010): 445–53.

10. Stellman et al., as cited in Institute of Medicine, *Veterans and Agent Orange: Update 2008* (Washington, DC: National Academies Press, 2009)

11. Ibid.

12. Agency for Toxic Substances and Disease Registry, "Public Health Statement: Chlorinated Dibenzo-P-Dioxins (CDDs)," December 1998, https://www.atsdr.cdc.gov/ToxProfiles/tp104-c1-b.pdf .

Chapter 6
Rainbow Herbicides

No one knows who first dubbed the destructive herbicides used in Vietnam "rainbow," but it's definitely a contradiction in terms. Rainbows are dazzling and beautiful. They have always been used to symbolize hope and a brighter future. They are a fascinatingly colorful natural phenomenon—stunning arches of colors in the sky caused by a combination of pure sunlight and divine rain. In fact, some ancient people even believed that a rainbow was a bridge between heaven and earth. But the herbicides used in Vietnam fall far short of that description. Sunset herbicides would be more apropos.

While very stunning to look at, rainbows can be caused not only by rain but by mist, spray, or any type of airborne droplets. I'm pretty sure that many veterans who were close enough to pesticide spraying operations saw rainbows in the sky as the noxious mist floated in the atmosphere. The sight of an "herbicide-bow" would not be so glorious to behold. While it might still look beautiful, it would signify the end of a brighter future and the beginning of a nightmare.

The herbicides used in Vietnam, like us, were all unique. They were an assortment of organically complicated multifaceted toxic chemicals, with Agent Orange being the most heavily used of the group. Agent Orange's complexity was due not only to its harmful compounds and lethal contaminant but also because of the different chemical formulations used in its production through the years. Similarly, the lack of contemporary statistical parameters and information regarding its manufacture and spraying and our contacts with it also added to the intricacy and diverse health problems it would ultimately produce.

Agent Orange I (used 1965–1970)

Active ingredients: 2-ethylhexyl ester of 2,4,5-trichloro-phenoxyacetic acid, an ester of 2,4-dichlorophenoxyacetic acid
Contaminated with: Dioxin-2,3,7,8-TCDD, percentage variable
Diluting agent: (1:10–20) parts of JP-4 or diesel fuel (contaminated fuel was acceptable, as per military policy)
Inert ingredients: percentage unknown

Agent Orange II (used after 1968)

Active ingredients: n-butyl ester 2,4-D and isooctyl ester 2,4,5-T
Contaminated with: Dioxin-2,3,7,8-TCDD, percentage variable
Diluting agent: (1:10–20) parts of JP-4 or diesel fuel
(polluted fuel acceptable, as per military policy)
Inert ingredients: percentage unknown

Agent Orange III (Time frame unknown)

Active ingredients: n-butyl ester 2,4-D and n-butyl ester 2,4,5-T
Contaminated with: dioxin-2,3,7,8-TCDD, percentage variable
Diluting agent: (1:10–20) parts of JP-4 or diesel fuel (polluted fuel acceptable, as per military policy)
Inert ingredients: percentage unknown

Enhanced Agent Orange (Time frame unknown)

(a.k.a. DOW Herbicide M-3393 or Super Orange, a standardized Agent Orange mixture of 2,4-D and 2,4,5-T combined with an oil-based mixture of picloram)

Contaminated with: dioxin 2,3,7,8-TCDD, hexachlorobenzene (HCB), and nitrosamines, all percentages unknown
Diluting agent: (1:10–20) parts of JP-4 or diesel fuel (polluted fuel acceptable, as per military policy)
Inert ingredients: percentage unknown

As noted, there were at least four different chemical designs for the herbicide we called Agent Orange. Consequently, as each formulation had an altered chemical configuration, each developed creation would have had a slightly different interaction with the veterans exposed. The icing on the Agent Orange cupcake was the allowed use—by military orders—of contaminated fuels to dilute any type of oil-based herbicide. According to military research, the chief pollutants of jet fuel—singly or in combination—were established to be water, dirt (especially iron rust), surfactants (unknown ingredients to improve efficiency), and microorganisms (really, as if JP-4 wasn't nasty enough).[1]

Bizarre Complications of Jet Fuel (JP-4)

JP-4 (a blend of 50 percent kerosene and 50 percent gasoline) was widely used in Vietnam, both as jet fuel and to dilute oil-based pesticides, such as Agent Orange and DDT. However, JP-4 also contained an intricate hazardous group of chemicals known by the acronym BTEX. According to the Environment Registry, "BTEX is the term used for benzene, toluene, ethyl-benzene, and xylene. Volatile aromatic compounds typically found in petroleum product, such as gasoline and diesel fuel."[2]

All the chemicals in BTEX are incredibly toxic to humans. In fact, it has even become a major environmental problem right here in the United States, especially around army and air force bases similar to the ones we had in the Cam Ranh area. While those of us working in motor-pool areas had heavier exposure to gasoline and diesel fuel, we all had a high probability of being contaminated with BTEX Just by living in the country.

Now, to give you an idea of just how much fuel was used in Vietnam, here are some petroleum-storage statistics noted in US Army Support Command reports for just the first quarter of 1968 for only the Cam Ranh Depot:

Page 25—Period summary—(in gallons)

CAM RANH	RECEIPTS	ISSUES	PRODUCT HANDLED
JP4	39,976,000	35,172,000	75,148,000
Avgas	4,088,000	3,644,700	7,752,000
Mo-gas	3,357,000	3,924,000	7,231,000
Diesel	6,561,000	7,170,000	13,731,000 [3]

Our exposure to BTEX could have occurred by merely consuming contaminated food or water, by inhalation, or even by absorption through our skin. Inhalation and absorption can occur from aerial spraying, while dispensing fuels, or while showering with tainted water. Severe exposure to high levels of BTEX has been medically associated with skin and sensory irritation, central nervous system depression, and adverse effects on the respiratory system. Thus, our extensive exposure to low levels of BTEX all by itself would have been injurious.[4]

While all the various chemicals contained in BTEX are dangerous, the most significant health hazard and medical concerns come from its primary component, benzene. Benzene is an exceptionally menacing substance found not only in JP-4 but in almost every kind of petroleum fuel and product used around the world.[5] In fact, the US Department of Health and Human Services released a public health statement for benzene in August 2007, advising that long-term exposure to low levels of benzene can cause leukemia, as well as damage the immune system and trigger other types of cancer.[6]

We have confirmation from military records that the use and storage of petroleum products were extensive at the Cam Ranh army and air force bases. We know that both kerosene and JP-4 were widely used to dilute oil-based pesticides, while diesel and gasoline fuels were widely used at our base motor pool. Thus, BTEX was capable of contaminating us directly as well as by leaching into the soil and—over time—polluting our water supply. However, no studies, to my knowledge, were ever done on the possible infiltration of our water supply by BTEX or any of the other toxic chemicals noted in this book.[7]

Then again, given the enormous amount of toxic military pesticides and chemicals used in and around the Cam Ranh area, the chances are, almost without question, that our potable water was contaminated to some extent with one or more of the known pesticide chemicals or their primary metabolites. The widespread aerial application of herbicides and insecticides would almost certainly have polluted the regular open-surface water sources from which livestock drank, the water we used for our showers, and even the ocean's surface.[8]

Also contained in JP-4 are what scientists call polycyclic aromatic hydrocarbons, or PAHs. This chemical class occurs naturally in any type of fossil fuel and consists of well over one hundred different chemicals banned together under one banner—PAH. Many PAHs have toxic, mutagenic, and/or carcinogenic properties, and they are readily absorbed into our bodies through the gastrointestinal tract. Upon ingestion or absorption, they are rapidly distributed to a wide variety of tissues, with an apparent predisposition for concentration in fat cells. The long-term health effects of exposure to PAHs can include cataracts, kidney and liver damage, and jaundice. While long-term contact with low levels of PAHs has been demonstrated to cause cancer in laboratory animals, studies of civilian employees exposed to mixtures of PAHs and other compounds have also noted a higher danger of skin, lung, bladder, and gastrointestinal cancers. The data provided by these studies are limited because, like us, these civilian employees were also exposed to other potentially cancer-causing chemicals besides PAHs. So, picking just one is hard to do.

Thus far, we have just touched the surface of Agent Orange and its toxic jet fuel diluent, and we have already found a horde of chemicals that were contained in the tactical herbicide used in real life. Regrettably, that was only the tiny tip of the pesticide iceberg. To further confuse an already complex situation, each of the various detrimental chemicals used in Vietnam presented different toxic issues and health problems for the personnel exposed to them. Unfortunately, to understand how all the chemicals and circumstances were interwoven, we must first deconstruct the complex cluster of compounds and look at each of the most significant components individually.

I know, it's a bummer—and probably a little tedious too. Even so, we must bite the bullet and examine each chemical substance separately before we can even begin to sensibly and realistically understand or try to determine what the combined health impacts of those substances were on the people exposed to them. So let's get started. First up: 2,4-dichlorophenoxyacetic acid (2,4-D), a genuinely intimidating chemical name that will cause even nastier health problems just by itself.

In the publication *Inert Ingredients Overview and Guidance*, the EPA identifies nearly three thousand substances that are being used as inert ingredients in pesticides here in the United States. Each of these three thousand chemicals has a wide and variable range of toxicity.[9]

The following is a list of just a few of the inert chemical substances that have been used in conjunction with the 2,4-D in herbicide formulations. Keep in mind, as you look over the list and associated health concerns, that 2,4-D was only one of the organic chemicals contained in Agent Orange and Agent White:

Amorphous silica: Diarrhea and obstruction of lung blood vessels

Aromatic solvent naphtha: Reduced fertility, reduced litter size, and reduced growth

Attapulgite-type clay: Cancer and tumors

1,2-Benzisothiazolinone-3-one: Genetic damage to human cells

n-Butyl alcohol: Severe eye irritation, genetic damage in hamsters, reduced fertility, and developmental abnormalities

Butyl Cellosolve: Severe eye irritation, genetic damage as demonstrated in bacterial tests, and sperm damage

Ethylenediaminetetraacetic acid: Genetic damage in laboratory animals

Ethylene glycol: Genetic damage in laboratory animals and in human cells, developmental abnormalities, diarrhea, nausea, headache, and reduced liver function

8-Hydroxyquinoline sulfite: Genetic damage, as demonstrated in bacterial tests and in human cells

Kerosene/Fuel oil No. 1: Severe skin irritation, genetic damage, as shown in bacterial tests, coughing, nausea, depressed activity, muscle weakness, anemia, and skin inflammation

Octylphenol polyethoxylated: Genetic damage in human cells and laboratory animals, developmental abnormalities, and skin inflammation

Propylene glycol: Genetic damage in laboratory animals, reduced fertility, high blood sugar levels, anemia, and tumors

Quartz silica: Genetic damage in laboratory animals and human cells, cancer, lung fibrosis, and diarrhea

Sodium benzoate: Genetic damage in laboratory animals and in human cells, developmental defects

Titanium dioxide: Genetic damage in laboratory animals, cancer, tumors, and diarrhea[9]

The above-listed chemicals are just some of the everyday inert ingredients investigators have discovered to be used in conjunction with 2,4-D. While the EPA has not determined the ability of purified 2,4-D to cause cancer, many studies have found that exposure to 2,4-D—the finished product—used in realistic applications has the potential to significantly increase the risk of lymphoma, which is a type of cancer that affects the immune system. Moreover, 2,4-D has continually been classified as a powerful endocrine-system disrupter, with some studies showing that human cells cultured in the presence of 2,4-D undergo significant chromosomal or genetic damage.[10]

The real-world application of 2,4-D has been studied widely, and countless peer-reviewed reports have confirmed a varied range of detrimental effects to the health of humans and animals exposed to it—numerous negative influences ranging from liver toxicity to non-Hodgkin's lymphoma. In addition, other studies have found 2,4-D to be an irritant to the gastrointestinal tract, causing nausea, vomiting, and diarrhea.[11]

What's more, several other studies of 2,4-D have confirmed that it is very capable of substantially affecting the human immune system. One such study, conducted by Canadian researchers, found that low levels of 2,4-D reduced the activity of several human genes that produce proteins having crucial immune system functions.[12] Many adverse health impacts from long-term exposure to low levels of 2,4-D have also been confirmed by studies to include blood, liver, and kidney toxicity, while long-term exposure to high levels of 2,4-D can cause stupor, coma, coughing, burning sensations in the lungs, loss of muscular coordination, nausea, vomiting, and dizziness.[13]

Finally, to put the frosting on the toxic cake—as if all the cancers and illnesses weren't bad enough—2,4-D has also been proven to contain the same deadly 2,3,7,8-TCDD dioxin found in the chemical 2,4,5-T, albeit at lower levels. The simple truth is that as of 2005, the chemical 2,4-D was the seventh most significant source of dioxins in the United States, according to the EPA.[14]

Thus, the 2,4-D used in the real-life environment of Vietnam was an extremely injurious substance in its own right. This important fact is often obscured by our government and is typically overlooked by most researchers. As a matter of fact, the 2,3,7,8-TCDD component of Agent Orange is being shamefully used by our government to obscure or deny all the evidence regarding the toxicity of 2,4-D.[15]

Next Up: 2, 4, 5-Trichlorophenoxyacetic Acid

If you thought 2,4-D and its health impacts were worrisome, you're going to be shocked by what 2,4,5-T can produce. The chemical 2,4,5-T is the very same chemical that was being manufactured during the Seveso Incident in 1976. It is also a major component of Agent Orange, the one that was the most severely contaminated by TCDD.

While the dioxin family as a whole comprises some of the most toxicologically devastating and complex chemicals in the world, just how deadly is the dioxin known as 2,3,7,8-TCDD? It's so lethal that less than one microgram (0.000001 gram) is enough to kill an adult guinea pig.[16]

Happily, we are a lot bigger than guinea pigs, and our physiology is very different. Nonetheless, in spite of this unusually high toxicity, the US government sprayed herbicides containing not only 2,4,5-T but also 2,3,7,8-TCDD as late as April of 1970, when the US DOD had suspended their use.[17]

While most pesticide industry studies show that the purified version of 2,4,5-T has relatively low toxicity in animals, it's almost impossible to separate the impacts and health effects of the sanitized 2,4,5-T from its primary manufacturing contaminant that is found in the real-world pesticide sprayed into our environment.

Therefore, these chemicals (2,4,5-T and TCDD) will go together, hand-in-hand, in their practical applications within our surroundings. As you might suspect, many studies have shown that exposure to the chemical 2,4,5-T in formulated end products has the same impact as contact with its major impurity, dioxin-2,3,7,8-TCDD, as discovered by the Seveso Incident studies.

How Much Dioxin-2,3,7,8-TCDD Is Too Much?

One quadrillion is the number to remember about dioxin. While it's really hard to conceive of just how big the number one thousand trillion is in the first place, it's even harder to consider it regarding the smallness of dioxin impurities. Scientifically speaking, dioxin—the one we call 2,3,7,8-TCDD—is toxicologically an extraordinarily problematic and biologically destructive organic chemical. So the following quote from the EPA about water contamination of 2,3,7,8-TCDD should come as no surprise to you, although the minuteness of the amount might:

> EPA has set this level of protection based on the best available science to **prevent potential health problems**. EPA has set an enforceable regulation for dioxin, called a maximum contaminant level (MCL), at 0.00000003 mg/L or 30 parts per quadrillion.[18]

Just as a point for your reflection, thirty parts per quadrillion is equivalent to thirty drops of ink placed into a fifty million cubic meter (13.2 billion gallons) container of water. This cube-shaped basin of water would measure roughly 403 yards on all sides. The actual height of that cube would be a little less than the height of the Empire State Building's 120 stories. That's a lot of water and a minuscule amount of dioxin.

The EPA is required by law to determine the highest level of toxic contaminants allowed to be present in drinking water at which no adverse health effects are likely to occur. But there is absolutely no known safe level for dioxin 2,3,7,8-TCDD. There is only our scientific limitation in measuring such microscopic amounts. So while the goal for dioxin is zero, the EPA has still set an enforceable MCL of 0.00000003 mg/L.[19]

This figure, thirty parts per one thousand trillion, should be kept uppermost in your mind whenever you hear our leaders or corporate officials talk about the "safe" levels of dioxin that offer no probability of causing long-term health problems in humans. No matter how you try to spin it, a fact is still a fact. There are no safe levels of dioxin, period.

The bottom line is that while many chemical company studies may have shown that 2,4,5-T's toxicity can be significantly lower in its sanitized form, its harmfulness increases dramatically and exponentially when it's combined with other chemicals and when the level of its deadly byproduct, 2,3,7,8-TCDD, rises during quick high-heat production.

The Agent Orange sprayed in Vietnam contained, at the very least, the following known ingredients: 2,4-D and 2,4,5-T as well as dioxin 2,3,7,8-TCDD, and it was diluted with JP-4 or kerosene containing BTEX and PAHs.[20]

Then There Was Agent White, aka Tordon 101

I hope you didn't think we were done with herbicides yet. It's a pretty big iceberg, and we still have one more major herbicide left. Next up is Agent White, which was actually called Tordon 101; we veterans gave Agent White its name because of the white stripe on the drums, just like we gave Agent Orange its name because of the orange stripe.

Tordon 101 was a mixture of picloram (4-amino-3,5,6-trichloropicolinic acid) and 2,4-D, in a ratio of roughly 1:4 (one part picloram to four parts 2,4-D). Picloram all by itself is extremely persistent in our environment and has an extremely high probability of leaching into groundwater. It's highly soluble in water and will move through earth/soil like a hot knife through butter.[21]

Agent White belongs to the pyridine carboxylic acid family of herbicides, and if it was diluted and used according to Military Assistance Command, Vietnam (MACV) instructions, it would have been in a 1:50 ratio of clear water, contaminated water not being acceptable. It was the second most heavily sprayed tactical herbicide during the Vietnam War.

Tragically, Agent White gets very little attention, even though it is an herbicide just as hazardous and health damaging as the number-one herbicide, Agent Orange. So while I have never believed what Walter Hagen and President Trump said—"No one remembers who came in second"—I might have to reconsider in the case of Agent White.

All the same, not listed as ingredients of Agent White are the following inert and side-reaction chemicals:

1. Hexachlorobenzene (HCB), a highly toxic manufacturing contaminant

72

2. The solvent triisopropanolamine

3. Nitrosamine compounds, which are generally considered by medical science to be carcinogenic

4. Any of the other previously noted substances associated with 2,4-D, (i.e., TCDD)

All the noted chemicals were contained in the Agent White manufactured for our military, but they were not included in the official list of ingredients. The reason is simple: they were all considered—by both the manufacturer and our government—to be "inert" or "trade secret" ingredients or manufacturing/storage side reaction impurities.[22]

Even so, tests results submitted by Dow Chemical to secure the registration of Tordon 101 concluded that the purified, cleaned-up version of picloram, no matter what the form, had a low toxicity rating. Conversely, experiments on picloram completed by EPA researchers showed it to be more toxic than was revealed by Dow's testing.[23]

Even more important is the fact that according to the EPA, the maximum contaminant level goal (MCLG) for picloram in drinking water is only 0.5 mg/L or 500 ppb. Once again, for reference, 1.7 US tablespoons of ink in a tanker truck holding roughly 9,100 gallons of water would create an ink concentration of approximately 500 ppb.

It is also worth mentioning that the EPA's MCLGs are set as close to health goals as possible while considering cost, benefits, and the ability of public water systems to detect and remove contaminants using suitable treatment technologies.

While the picloram contained in Agent White was quite hazardous, the 2,4-D and the HCB present in the finished product made it even more dangerous. Deceptively, while 2,4-D was noted as an active ingredient, the HCB content of Agent White was hidden by merely classifying it as a contaminant or side reaction product. Thus, in accordance with pesticide laws, because HCB was not purposely added to the pesticide formulation, it is not considered an active or even an inert component of Agent White but merely an impurity. This implies that the producer's scientists had no idea that the mixed chemicals would react to produce HCB.[24]

In fact, as I investigated the complete documentation submitted to support the registration eligibility of Tordon 101, I discovered that a significant number of supporting studies were unpublished. As you may or may not know, unpublished studies are not subject to peer review for inferior scientific procedures or conclusions (i.e., no fact-checking).[25]

The Tordon 101/Agent White used in Vietnam was and continues to be a proprietary product of the Dow Chemical Company. Dow claimed, based on its own tests, that the pure active ingredients—picloram and 2,4-D—in its products had a low toxicity rating. But Dow's claim was apparently not for the finished product, which was laced with HCB and TCDD.

Nevertheless, in 1985, when Dow Chemical had to recertify Agent White/Tordon 101, they lost the registration and license for this herbicide. The primary reason was that their creation was found to be severely contaminated by hexachlorobenzene. Thus, Dow would not be able to reclaim and sell their Tordon 101 products until mid-1988.[26]

Unfortunately, we'll probably never know the extent to which the Agent White used in South Vietnam was contaminated. What we do know is that according to the EPA's Red Facts, for Dow Chemical to regain the registration of its Agent White products, it had to reduce the contamination to "less than 200 parts per million of HCB and less than 1 percent nitrosamines."[27] While Dow did comply with the EPA rulings, there is still no way of knowing how severely the real Agent White atomized and sprayed in Vietnam was contaminated with these "secret" chemical substances and impurities.

Likewise, we don't know what the synergic interactions would have been or what the negative health impacts were on veterans. What we do know is that the use of HCB has been expressly banned in at least 120 nations around the world because of its extremely negative health impacts on humans.[28]

The bottom line is this: all the Agent White used in Vietnam was severely contaminated with HCB and tainted to some extent with the impurity dioxin-TCDD via 2,4-D. As a result, our governmental and military leaders should have known the toxicity and exposure consequences of Agent White (in addition to those of Agent Orange) before they allowed its deployment in the immediate areas of US bases and personnel.

Lastly, while the chemical triisopropanolamine is considered to be less toxic than the other ingredients in Agent White, no studies have been conducted to date on the interaction of this particular chemical ingredient with respect to its use in combination with others, such as Agent Orange, malathion, and DDT or with other chemicals, such as BTEX and PAHs.

74

There is very little doubt that the unlisted chemicals in Agent White made it an even more dangerous pesticide—an herbicide that should never have been used in Vietnam without first being tested and studied to determine its real-world health consequences for military personnel.

Toxicity of Organochlorine Pesticides

We're almost finished with Agent White, and the last substance I want to discuss is hexachlorobenzene. HCB is a persistent chlorinated organic chemical in the same class of pesticides as DDT and its sister compound, dichloro-diphenyl-dichloro-ethylene (DDE). HCB is listed by the World Health Organization as extremely injurious. Consequently, its global use has been banned since 1965. HCB causes a wide range of effects once inside the human body. However, for most of these effects to occur, it appears that HCB must be altered by our bodies into its primary metabolites.[29]

When the EPA reviewed hexachlorobenzene as part of its regular six-year review process, it determined that the appropriate MCLG was zero, with an MCL of just a scanty 0.001 mg/L (1 ppb). This small amount was set by the EPA to be protective of human health.[30]

Research has confirmed that Agent White was severely contaminated with HCB. We also know that picloram, 2,4-D, and HCB are individually suspected of triggering dozens of diseases and organ dysfunctions. In point of fact, the Comparative Toxic Genomics Database (CTD) website, (http://ctdbase.org) lists a staggering number of possible illnesses, conditions, and dysfunctions associated with just these few chemicals.[31]

If we were to use the CTD database to investigate the probable negative health impacts caused by the interactions of HCB, picloram, triisopropanolamine, and nitrosamines—only the known chemicals—we would discover a wide range of devastating health impacts just from Agent White by itself.

With the health impacts almost certainly produced by Agent White still in your mind, consider what happens to the body and health of veterans when you add in exposure to all the chemical ingredients found in Agent Orange and JP-4 jet fuel. Now, atomize and combine all of them, and what do you get? You get adverse health impacts that would be toxicologically devastating to the biological systems of any human exposed to just these two pesticides.

The damage that these two color-coded herbicides can produce collectively to our cellular, hormonal, and genetic makeup is mind-boggling. Still, when the exposures are combined, the cellular and genetic alterations that are scientifically and medically known to occur within the various hormonal and biological systems of humans and animals will be, statistically speaking, astronomical. Not only is the mere environmental exposure to herbicides a problem, but there is also what scientists call internal exposure that must be considered. As if direct contact wasn't bad enough, research advises that anyone exposed to organic chemicals, such as TCDD, HCB, and the rest of the nasty gang, can store those substances within the cellular structure of their bodies—mostly in fat cells. Thus, our bodies will continue to experience the impacts of internal pesticide exposure for years, even after we leave our contaminated environment and direct contact has stopped.[32]

Military records have confirmed that almost six million gallons of Agent White concentrate or approximately three hundred million gallons of end product—if correctly diluted—was used in Vietnam. We know that more than 77,215 gallons of Agent White concentrate—almost 4 million gallons diluted—was sprayed on Khanh Hoa Province, where Cam Ranh is located. Also near Cam Ranh is Ninh Thuan Province, where roughly 2,075 gallons of concentrated Agent White was also sprayed. Of those two totals, at the very least, more than 1,320 gallons of Agent White concentrate—66,000 gallons diluted—and very possibly even more because of a phenomenon called pesticide drifting or just simply mist drifts—were sprayed on Cam Ranh directly.[33]

While Agent Orange and Agent White were only two of the herbicides sprayed in Vietnam, they were the most prevalent. All the same, that brings us around to the inevitable issue or question that must be asked and answered: Should herbicides have been used in Vietnam to begin with, and were there any alternatives to their use?

I'm pretty sure that the debate as to whether or not herbicides should have been used during the war will continue for decades to come. So even as the controversy rages on, the simple truth is there were many other nontoxic strategies that could have been employed. In fact, as herbicides were being sprayed throughout Vietnam, our military was also using at least two types of mobile equipment that were very capable of mowing down vast areas of jungle.

The first massive vehicle was initially designed for US logging firms and was tenderly dubbed the Tree Crusher.[34] This massive sixty-ton vehicle consisted of two sharp five-bladed roller-type wheels in front of the driver's cab and one in the rear to steer. As the monster rototiller moved forward, it would use the sheer force of its engine to knock over trees and mow down dense brush that stood in its path. The sharp five-bladed rollers would then break up the trees and brush as the massive five-bladed wheels rolled over them. The second vehicle was the Rome plow, also called hog jaws by military engineers.[35]

Hog jaws was the US military's version of a large D7E bulldozer outfitted with a 2.5-ton blade. It also had a steel-clad protective cage mounted on top to shield the operator. This eighteen-ton monster was initially used by US forest firefighters to create firebreaks but eventually found its way to Vietnam, making its first appearance in 1966. Upon its arrival, the "jungle eater" was put right to work clearing the dense jungle and forest areas in Vietnam.

Hog jaws was apparently so successful that by January 1967, the US Army launched a top-secret operation called Cedar Falls. The stated purpose of the mission was to destroy a Viet Cong stronghold which had been dubbed the Iron Triangle. Fundamentally, the zone to be destroyed was actually just a triangular-shaped twenty-six-thousand-acre expanse used by the VC as a staging area to launch brutal attacks on the city of Saigon.

Now I could go on and on about Operation Cedar Falls, which was followed by Operation Junction City, but those battles are already part of the historical records of the war, and there is no need to belabor them past the point that I'm trying to make. The truth is that there were other non-toxic methods the military could have used to clear strategic areas of Vietnam without lambasting everyone with harmful herbicides. Insects and insecticides, however, were a whole different ball game.

Endnotes—Chapter 6

1. Military Research on Jet Fuel Contamination By Donald B. Brooks May 13, 1963

2. Environment Registry-Item: benzene, toluene, ethylbenzene, xylene http://environment.data.gov.au/def/object/BTEX

3. Operational Report - Lessons Learned, Headquarters, US Army Support Command Cam Ranh Bay, Period Ending 30 April 1968 – (1st Quarter 1968)

4. Fact Sheet BTEX - Maryland Department of the Environment

5. Science Corps. - Veterans health - Health Hazards of Chemicals Commonly Used on Military Bases

6. Agency for Toxic Substances and Disease Registry (ATSDR), U.S. Department of Health and Human Services - *Public Health Statement for Benzene* – August 2007 - CAS#: 71-43-2

7. The United States Environmental Protection Agency – EPA Superfund Record of Decision: Hill Air Force Base 9/29/1998

8. The National Academies of Science - Veterans and Agent Orange: Health Effects of Herbicides Used in Vietnam – Chapter 3 - The U.S. Military and the Herbicide Program in Vietnam.

9. The Journal Of Pesticide Reform Winter 2005- Inert Hazards in 2, 4-D herbicide

10. *Pesticides News*, September 1997, p. 20.

11. The National Academies of Science - *Veterans and Agent Orange: Update 2012 (2014)*

12. Journal Of Pesticide Reform - Winter 2005 • VOL. 25, NO. 4- updated 4/2006 – *Herbicide factsheet*: 2,4-D

13. National Pesticide Information Center - *2, 4-D Technical Fact Sheet*

14. Center for Food Safety - Parkinson's disease and 2, 4-D: A Summary of the Evidence May 2013 (updated May 2014).

15. Center for Food Safety - Parkinson's Disease and 2, 4-D: A

Summary of the Evidence May 2013 (updated May 2014)

16. Ibid.

17. The National Academies of Science - *Veterans and Agent Orange: Update 2012 (2014)*

18. EPA's Ground Water & Drinking Water - Fact Sheets: Dioxin (2,3,7,8-TCDD) - 4. What are EPA's drinking water regulations for dioxin?

19. EPA National Primary Drinking Water Regulations

20. National Pesticide Information Center - *2, 4-D Technical Fact Sheet*

21. Journal Of Pesticide Reform – Spring 1998 • VOL. 18,

22. Ibid

23. EPA R.E.D. Facts – Picloram (EPA-738-F-95-018) 1995

24. *Agent White a.k.a. Tordon101* - by Lieutenant Colonel Patrick H. Dockery (USAR, RET) 20 June 2000

25. Ibid

26. Environmental Protection Agency – *R.E.D. Facts Picloram* -EPA-738-F-95-018 August 1995

27. EPA Protocol for the Review of Existing National Primary Drinking Water Regulations One PPB would equal one drop of water in a tanker truck holding 13,200 gallons.

28. Agent Orange Association of Canada - The Missing Part of the Equation – Hexachlorobenzene (HCB)

29. The National Center for Biotechnology Information (NCBI) - *The role of oxidative metabolism in hexachlorobenzene-induced porphyria and thyroid hormone homeostasis:* a comparison with pentachlorobenzene in a 13-week feeding study.

30. EPA Protocol for the Review of Existing National Primary Drinking Water Regulations One PPB would equal one drop of water in a tanker truck holding 13,200 gallons.

31. The database is maintained by the Department of Biological Sciences at North Carolina State University and the Department of Bioinformatics at MDI Biological Laboratory.

32. National Academies of Science-*Veterans and Agent Orange: Update 2012 (2014)* Pg. 70.

33. *H.E.R.B. Tapes: Defoliation Missions in South Vietnam, 1965-1971*. Data by Province.- Alvin L. Young Collection on Agent Orange Container List - document # 00108). Also the *Chicago Tribune's interactive map* – This map is based on the Herbicide Exposure Assessment - Vietnam database developed by Dr. Jeanne Stellman, professor emeritus at Columbia University's school of public health, and Columbia epidemiology professor Steven Stellman. According to this interactive map and HERB tapes, at the very least, Cam Ranh and/or the close surrounding areas were sprayed on or about the following months and years with Agent Orange or Agent White:

1966 – September

1967 - June, July, August, and September

1968 – April, August, September, November, December

1969 – February, and April

The Chicago Tribune's interactive map also noted that the heaviest spraying of the herbicides AO, AW and AB occurred between 1967 and 1969. My tours of duty in Vietnam extended from September 1966 through June 1969. This time period was also the phase of the heaviest herbicide/insecticide spraying during the war according to military records and studies.

34. This Is The World's Biggest Tree Crusher! The Monstrous Le Tourneau G175! From Muscle Car Zone

35. Vietnam War Slang: A Dictionary on Historical Principles by Tom Dalzell

Chapter 7
Insecticides

Let's face it: mosquitos, flies, and other insects were regular aggressors in Vietnam and just part of everyday life there. It didn't matter how clean you kept your quarters—bugs were ever-present. While the military's strategy to use herbicides was very questionable, the use of insecticides was not. The military plan developed to keep pest-borne diseases to a minimum was an excellent tactic. Thus the extensive spraying of insecticides during the war was an easy strategy to justify.

Don't get me wrong: bugs were an ever-present problem, and bug sprays and repellents were needed. The problem came with the pesticide formulations being so cheaply and rapidly made by avaricious chemical companies. As a result, the finished and stored products were not only complicated bug killers but also extremely harmful to humans exposed to them. There is no doubt that the insecticides used in Vietnam could have been made safer and much less toxic, but they weren't.

Insecticides are an integral part of the pesticide family, and as the name implies, they are specially formulated to target and kill insects. Some bug sprays are designed to disrupt the nervous system of the creepy-crawlies, while others damage their exoskeletons or just repel them. While Agent Orange has received most, if not all, of the spotlight, the insecticides used in Vietnam were just as life altering and health damaging and, in some cases, even more harmful than the herbicides sprayed on us.

For the most part, the insecticides used in Vietnam fell into four broad classifications:

1. Organochlorine pesticides (e.g., DDT, dieldrin, and lindane), up-close insecticides used to treat anything from uniforms to bedding and fleas to body lice

2. DEET, personal insect repellents used on our skin

3. Organophosphorus pesticides (e.g., malathion)

4. Pyrethroids, which are pyrethrum-like pesticides (e.g., permethrin)[1]

Our military has loudly proclaimed and credited the extensive use of insect-killing sprays, powders, and personal bug repellents for keeping rates of pest-borne diseases low during the Vietnam War. However, they have remained silent on the toxic aspects of those very same sprays and powders. Since there were no studies conducted before or after the Vietnam War on the insecticides used there, we must turn instead to Gulf War studies for information. The war was different, but the bug sprays were almost exactly the same.

The results from studies on the insecticides used by the US military during the Gulf War have indicated linkages between exposure to organophosphate pesticides and neurocognitive deficits and neuroendocrine alterations in Gulf War veterans. As a matter of fact, a 2008 report determined that the combined evidence supports a reliable and convincing case that pesticide use during the Gulf War was causally linked with Gulf War illness.[2] While the Gulf War studies on organophosphate pesticides are disturbing on many levels, most troubling is the fact that they were also used in Vietnam in addition to organochloride insecticides. In fact, one of the first widely used insecticides in Vietnam was the now-infamous organochloride pesticide known as DDT.

DDT was used in combination with pyrethrum and cyclohexanone, an auxiliary solvent. As an environmental chemical, DDT is converted into several breakdown products or metabolites, one of which is dichloro-diphenyl-dichloro-ethylene (DDE). DDT and DDE are both stored in the fatty tissues of anyone exposed to them. Both DDT and DDE are very hazardous all by themselves, while the inert ingredient pyrethrum has limited toxicity at low dosages, and the solvent cyclohexanone has low to moderate toxicity, at least according to the manufacturers.[3]

The adverse environmental and health impacts posed by DDT have been widely publicized, due in large part to the firestorm sparked by the release of Rachel Carson's book *Silent Spring* in September of 1962. Since then, the use of DDT has been banned internationally, and the health problems associated with it have been scientifically well established. I'm pretty sure by now everyone has heard or read about the evils of DDT, so I will just note it is an endocrine disruptor and a cancer-causing agent.

While DDT was used at a distance in Vietnam, powder products such as dieldrin and lindane were used up close for bedbugs, fleas and body lice. Even more up close and personal was an insect repellant that we applied directly to our exposed skin. The repellant was called N,N-diethyl-meta-toluamide, or DEET for short. We used the stuff on a daily basis and applied it quite liberally. Nonetheless, even this supposedly innocuous insect repellant had some pretty severe concerns associated with its use— in particular, the interactions it had with the many other chemicals contained in the pesticides sprayed on us veterans.

Over the intervening years, numerous scientific studies have demonstrated that some of the actual ailments related to the use of DEET include contact dermatitis and eye irritation. Despite the fact that DEET causes only minor health issues, there is still a major problem with its use. The fact that DEET is readily absorbed through both our stomach and skin makes it an essential factor in assisting other, more toxic insecticides and herbicides to be effectively absorbed into the human body.[4]

DEET's synergistic interactions with other pesticides have been very well documented. One instance is a 2001 study that showed a combination of DEET and permethrin—another chemical often sprayed for mosquito control—contributed to motor defects and memory dysfunction in humans. According to scientific research, most of the health impacts of DEET arise from its interaction with other, more toxic chemicals and its ability to penetrate the skin and stomach barriers.[5]

The next insecticide even more widely used during the Vietnam War was malathion. This pesticide is in the organophosphate family and was also used heavily during the Gulf War, as already noted. Operation Fly Swatter and other sources detail the spraying of tactical malathion over US military cantonment areas of Vietnam every nine to eleven days. Thus, personnel stationed in those designated areas were regularly sprayed and exposed to malathion and all its known and still-unknown compounds on a recurring basis.[6]

Unfortunately, once again, thanks to secrecy and poor record keeping, information is limited on the areas sprayed, as well as on the quantities of malathion, DDT, dieldrin, DEET, pyrethrum, lindane, and all the other harmful organophosphate and organochloride pesticides used during the war. Unluckily for us, as the mosquito population grew more resistance to DDT, our military, in a continuing attempt to reduce mosquito-borne diseases, eventually turned to malathion as their full-time, go-to bug spray. All the same, it wasn't until October 1966 that a Ranch Hand UC-123 herbicide defoliation aircraft would be reconfigured to spray malathion.[7]

In 1978, a US Defense Department official was quoted as having said that malathion and DDT were the other principal pesticides used in Vietnam, and they were used throughout the war for mosquito control: "Malathion was sprayed by aircraft, and DDT was applied by backpack and paintbrush." The official went on to say, "No information is readily available on the quantities used in Vietnam."[8] But just what is malathion, and what health issues does it produce?

Purified malathion is essentially a nerve poison that acts by inhibiting acetylcholinesterase enzyme receptors. Studies have shown that long-lasting polyneuropathy and sensory damage have been reported in people exposed to it, as well as behavioral changes. As a result of both animal testing and real-life spraying experiences, malathion (the technical or finished products) has been scientifically established to affect not only the central nervous system of mammals but also the immune system, adrenal glands, liver, and blood.[9]

While malathion is an insecticide, the primary formulation process is no different than that of an herbicide. It contains active, secret, and inert ingredients, plus side product contaminants. Formulated malathion (the kind we were exposed to) contains many unlisted chemicals that increase the toxicity of malathion well beyond that of its refined chemical version. These pollutants occur as side reactions during and after the production process. Their development and concentration are directly affected by the manufacturer's secret inert ingredients as well as by storage conditions. These impurities and inert components will unavoidably increase the toxicity of the sanitized laboratory version of malathion many times over.[10]

The following are some of the impurities that studies have found in the finished formulated malathion product (DLCs are presented in **boldface**):

1. Diethyl fumarate 0.90%

2. Diethyl hydroxysuccinate 0.05%

3. O,O-Dimethyl-phosphorothioate 0.05%

4. O,O,O-Trimethyl-phosphorothioate 0.45%

5. O,O,S-Trimethyl-phosphorodithioate 1.20%

6. Ethyl nitrate 0.03%

7. Diethyl mercaptosuccinate 0.15%

8. Isomalathion 0.20%

9. S-(1-Carbomethoxy-2-carbethoxy) methyl-O,O-dimethyl phosphorodithioate 0.60%

10. Bis-(O,O)-dimethyl-thiomophosphoryl-sulfide 0.30%

11. Diethyl methylthio succinate 1.00%

12. S-Ethyl-O,O-dimethyl-phosphorodithioate 0.30%

13. S-(1,2)-Dicarbethoxyl ethyl-O,O dimethyl phosphorothioate 0.10%

14. Diethyl ethylthiosuccinate 0.10%

15. Acidity as sulfuric acid (H_2SO_4) 0.05%

To date, there is no requirement to place or acknowledge any of the above substances as ingredients on malathion labeling; nevertheless, several harmful chemicals were contained in the finished insecticide used in the real-world environment of Vietnam.[11]

While most of the above ingredients are injurious in varying degrees, one contaminant stands out from the rest, requiring special attention. It is the storage impurity O,S,S-trimethyl phosphorodithioate (now, that's a mouthful), or OSS-TMP for short. Numerous studies have confirmed that just this one contaminant, OSS-TMP, is roughly five hundred times more toxic than the sanitized lab version of malathion.

Scientific and medical studies have demonstrated that malathion impurities will increase in volume during its storage, especially three to six months after manufacture, thus making the finished and stored insecticide far more injurious to people than it was when it was first manufactured. OSS-TMP and all the other impurities have been scientifically verified to increase more rapidly when malathion is exposed to high temperatures in the area of one hundred degrees.[12]

As a result, unless the military used the insecticide very quickly and kept the drums in an air-conditioned building, the pesticide would have been exposed to temperatures hovering in the hundred-degree range quite often. The resulting OSS-TMP—also being a DLC—would have added to the overall dioxin burden experienced by veterans.[13]

The multifaceted adverse health problems produced by exposures to just malathion include impaired fat metabolism, high blood-sugar levels, cellular insulin resistance, impaired glucose tolerance, asthma, hypersensitivity, allergies, memory loss, reduced speed of response to stimuli, reduced visual ability, altered and/or uncontrollable mood changes in general behavior, and decreased motor skills.[14]

The end result is, while the purified, "cleaned-up" lab version of malathion may have a low toxicity test rating, the heavy-duty, tactical-grade finished and stored product does not. The storage of the insecticide and its cheap and rapid manufacturing process increased the toxicity well beyond any acceptable limits, especially in the hot weather conditions of South Vietnam. These conditions, along with the presence of all the active, inert, and secret ingredients, made the actual real-life finished insecticide concentrate used in Vietnam much more health damaging, although considerably less expensive.

MACV Report on Spraying Statistics for Malathion – 1969

Now you might have thought obtaining information on all the pesticides used in Vietnam would be an easy task, but you would be wrong. It was not. Take for example the following quote from an official command health report currently housed in the US Department of Agriculture's National Agriculture Library Special Collections on Agent Orange. It speaks for itself, but I'm not really sure how it ended up in the custody of the USDA for storage and safekeeping. More importantly, it is the only report of its kind I could locate. Still, there should have been many more

86

declassified ones available to us by now:

> 7. Aerial Dispersal of Insecticides: (a) Air Force C-123 "Ranch Hand": Both C-123 aircraft continued in use. The target interval varied but is approaching the desired of 10 days. The following missions were flown in the month of June. The number accompanying the area indicates the number of missions if more than one: Phu Cat (3), Hue/Phu Bai (3), Due My (6), Da Nang (6), Ninh Hoa (3), Phan Rang (3), Chu Lai (2), Men Hoa (2), Ap Nam/Phuoc Tho (3), Tan Son Nhut (2), Long Binh (4), 199th SBs, **Cam Ranh (2)**, Dong Ha/Quang Trl (3), and Pleiku. A total of 15,775 gallons of 95% malathion was applied to the areas listed in the **46 missions completed.**[15] (emphasis added)

The above MACV report on mosquito spraying statistics and totals was only for one month, June 1969, and as noted, Cam Ranh was sprayed twice during that single month. Also, as pointed out in this report, the command was approaching the desired target of spraying selected areas every ten days. Thus, as documented, we know that Cam Ranh was being sprayed with malathion and all its contaminants on a regular basis from the beginning of the program in 1966 through at least the 1970s.

Further down in this same MACV report is an official recognition that unauthorized aerial sprayings of insecticides (DDT and malathion) from helicopters also occurred, which was inconsistent with the provisions of MACV Directive 40-10.[16] The report goes on to officially notify the United States Army Vietnam Command that 4,726 barrels (260,000 gallons) of 57J6 concentrated malathion appeared to have been used by unapproved sources.[17]

The report never clarifies that last statement. We don't know if it was because the concentrate was not diluted as per MACV instructions, or if the missions were not being recorded, or if the pesticide barrels were misappropriated. But the unaccounted-for loss of 260,000 gallons of a considerably harmful substance is not a good thing.

There is no question that, based on all the studies and information about malathion and DDT, they should be considered pesticides of interest by our government. But like many other pesticides, malathion and DDT have been lost in all the chaos surrounding Agent Orange and relentless governmental chants. As a consequence, malathion and DDT are not being acknowledged, discussed, or even considered by our legislative leaders. They are not being researched by any governmental agency, either by themselves or more importantly, in combination with all the other detrimental pesticides used there—exceptionally harmful substances such as those contained in Agent Orange, Agent White and BTEX.

What deserves even more attention is the fact that according to the EPA, all organophosphate pesticides act on the human body in a similar manner, and their effects are cumulative. Again, this raises health questions and concerns about Vietnam veterans' multiple exposures to malathion (roughly twice a month), combined with exposure to all the destructive chemicals contained in the other pesticides being sprayed into the environment over their tours of duty.[18]

The Metabolite Malaoxon

Sadly, we are not finished with malathion yet. We still have its metabolites[19] to consider. I'm not going to cover them all, just one—the one called malaoxon. Malaoxon is in the group of chemicals known as phosphorothioates.

Malaoxon is a member of a larger toxic family of substances that are scientifically considered to be dioxin-like compounds. Studies have shown that upon absorption, inhalation, or ingestion into the human body, malathion readily metabolizes to malaoxon, which is a substantially more toxic chemical. In fact, according to the *NPIC Fact Sheet*, malaoxon is considered to be twenty-two times more toxic than the parent compound (purified malathion) following acute dietary exposure and was found to be thirty-three times more toxic by all other exposure routes.[20]

The data collected by the EPA on the toxicology of malaoxon showed that any tactical-grade malathion (from spray runoff and mist drift) present in the downstream areas of water sources used for drinking would be altered. According to the EPA research, any wayward malathion existing in these raw water sources will form malaoxon during the chlorination process in water treatment facilities.[21]

Multiple exposures to tactical-grade malathion—just by itself—would have been detrimental enough to handle. However, when you logically consider the few pesticides I have documented, along with just the known chemicals they contained, you will begin to have a real appreciation and understanding of the entangled, multifaceted, nightmarish mess created during the war. The only way I can describe that chaos is to say that it was a genuine, shocking chemical soup.

Our combined exposures were so significant that keeping track of them or even accounting for all the numerous intricate combinations and interactions becomes a virtual impossibility. In reality, trying to individually substantiate or epidemiologically link to all, or even most, of the likely adverse health impacts produced by our environment in Vietnam is a real fool's errand without first studying veterans who served there. In fact, you might even say it's like a dog chasing its own tail. We'll never get there.

End Notes - Chapter 7

1. Institute of Medicine - Gulf War and Health: Volume 8: Update of Health Effects of Serving in the Gulf War - Appendix

2. Research Advisory Committee on Gulf War Veterans' Illnesses Gulf War Illness and the Health of Gulf War Veterans: Scientific Findings and recommendations Washington, D.C.: U.S. Government Printing Office, November 2008

3. The Centers for Disease Control and Prevention (CDC) - Environmental Health – Dichlorodiphenyltrichloroethane (DDT) Fact Sheet - November 2009

4. Naval Medical Field Research Laboratory List of Herbicides and Pesticides Used in 1968 by U.S. Armed Forces.

5. Mosquito Management and Insect-Borne Diseases by Beyond Pesticides

6. Operation FLYSWATTER: A War within a War by Paul F. Cecil, Sr. and Alvin L. Young

7. Ibid

8. Alvin L. Young Collection - document # 05245 - Herbicide "Agent Orange": Hearing before the Subcommittee on Medical Facilities and Benefits of the Committee on Veterans' Affairs, House of Representatives, Ninety-Fifth Congress, Second Session on Herbicide "Agent Orange," October 11, 1978 - Letter to the Honorable Ralph H. Metcalfe, August 16, 1978,

9. Beyond Pesticides – Chemical Watch Fact Sheet – Malathion Revised July 2000

10. Journal of Pesticide Reform, Volume 12, winter 1992 – *Insecticide fact Sheet – Malathion*

11. Dr. John Whitman Ray, N.D., M.D. (M.A.) 1993 - *12 Points on Malathion*

12. Journal of Agricultural Food Chemistry, 25(4):946–953, 1977 - Malathion Becomes More Toxic When Stored for 3 months or When Temperature Increases

13. Ibid.

14. *Malathion Medical Research* by Wayne Sinclair, M.D.

15. US Department of Agriculture National Agriculture Library – Special Collections - Agent Orange Item ID: 220 - *Command Health Report for June 1969* (RCS MED-3 (Rr-4) 28 July 1969

16. Ibid.

17. Ibid.

18. For additional studies on malathion, see "Intestinal Problems in Test Animals Exposed to Malathion," *Bulletin of Environmental Contamination Toxicology* 33 (1984): 289–94; Division of Toxicology and Physiology, University of California, Riverside, "Impurities in Malathion Found to Disable the Body's Natural Ability to Detoxify Malathion," *Toxicology and Applied Pharmacology* 49 (1979): 107–12; "Malathion Breaks Down into Even More Toxic Malaoxon," *Bulletin of Environmental Contamination Toxicology* 57 (1996): 705–12

19. Metabolites are the substances formed as part of the natural biochemical process in which a parent chemical is changed into different break-down compounds. For example, the insecticide malathion, once created and sprayed into our environment will start breaking down or changing its chemical structure to form monocarboxylic and dicarboxylic acid derivatives and malaoxon.

20. Journal of Pesticide Reform, Volume 12, Number 4, *Insecticide fact Sheet – Malathion*

21. Ibid.

Chapter 8
Military Base Chemicals

While I want to keep my focus on the issues of boots-on-the-ground veterans, I would be remiss as an investigator if I didn't discuss a piece of related information I discovered during my explorations. You might think, as I did, that my first encounter with hazardous chemicals occurred in Vietnam, but you would be wrong. Vietnam was where my most direct and harmful exposures occurred, but my first encounter with toxic pesticides, volatile organic compounds (VOCs), PCBs, and BTEX occurred right here in the United States.

My initial toxic chemical encounter was at Fort Dix, New Jersey, where I completed basic training. It seems that in 1982, New Jersey State investigators found groundwater contamination beneath one of the landfill sites being used by both Fort Dix and McGuire Air Force Base. The EPA investigated and discovered that Fort Dix had at least nine possible sites that were contaminated by numerous hazardous substances. One of the polluted sites was the sanitary landfill area, which has since been placed on the National Priorities (Superfund) List, due to organic solvents such as VOCs, BTEX, other organic PAHs, and heavy metals, including chromium, being found at the location.[1]

After graduating from basic training, I was assigned to Fort Devens, Massachusetts, for advanced training. As it turned out, the water and soil at Fort Devens were also contaminated with PCB, pesticides, VOCs, and petroleum products.[2] In fact, because of the many years of military operations, more than eighty areas were suspected of being possible hazardous waste sites. As a result of these alleged widespread polluted areas, the EPA had no choice but to add Devens to the agency's list of hazardous waste sites needing cleanup.[3] Unluckily for the people stationed there, after the army investigated, it turned out that there were actually 324 likely areas of contamination identified. Of those 324 sites, over 230 were eventually cleaned up, or at least the pollution was deemed to meet acceptable levels.[4]

In 1989, the EPA added Fort Devens to the National Priorities List of sites identified for long-term remedial response because of severe groundwater contamination. Today, Devens is mostly an inactive army base, and both Devens and Dix have undergone extensive cleanup with EPA oversight and taxpayer money. So while I had believed my toxic pesticide and chemical exposures were relegated exclusively to Vietnam, the reality proved to be entirely different.

Pesticide and toxic chemical pollution is a widespread, almost universal problem within numerous military bases here and abroad. Forts Devens and Dix are only two of the many bases that have been discovered to be contaminated by harmful organic chemicals and toxic heavy metals. Toxic exposures are a serious problem at many military bases, from Marine Corps Base Camp Lejeune[5] to Hill Air Force Base in Utah.

As a matter of fact, Hill AFB was also declared an EPA Superfund site after it was discovered to have BTEX contamination levels 1,250,000 times greater than those that the Occupational Safety and Health Administration (OSHA) consider harmful. The pollution was so expansive that BTEX levels of 5,000 mg/kg were found ninety-five feet below the surface.[6] Again, just as a reference, OSHA considers BTEX contamination above 0.004 mg/kg to be harmful to humans.

Hill AFB, regrettably, is not alone when it comes to BTEX contamination; Forts Dix and Devens, Luke Air Force Base, Mather Air Force Base, Travis Air Force Base, McClellan Air Force Base, and countless other military bases and fuel storage areas have been discovered to be extensively contaminated by BTEX in addition to many other harmful organic substances.[7]

Hazardous chemicals originate in many shapes and forms. They assault us with generic alphabet names—acronyms such as BTEX, PCP, and VOC. They come disguised in the form of supposedly beneficial herbicides and insecticides and other allegedly helpful chemicals that, when deployed into our environment, turn out to be health-damaging substances, especially when we're exposed to several different materials at the same time.

Our toxic chemical contacts in the military were not isolated single incidents relegated only to Vietnam. Instead, they were a string of events marked by combinations of daily exposures to numerous and diverse hazardous substances. In fact, many scientists have been arguing for years that toxic exposures to materials such as herbicides, insecticides, BTEX, PCP, and VOCs should be measured as they would typically occur in our environment and in combination with one another, but oddly, our government officials appear to lack the collective will to make that type of testing an actual reality.

Consequently, as in any proper investigation, the question of "Why?" must be answered, and this lack of a collective will does have a why. Inappropriately, once again, it's about the money. What our government scientists and chemical companies know—as a certainty—is that every toxic organic chemical we are exposed to has a maximum dose that will be 100 percent lethal (LD_{100}) to anyone exposed. Likewise, an LD_{50} rating means that at a specific dose, 50 percent of subjects exposed to the studied chemical will die. But what happens when the amounts get into the microscopic range such as with veterans' exposure to low levels of dioxin, HCB, and the like?

If the dosage is minuscule enough, nothing happens, or there is no observed adverse effect level (NOAEL), but there is a point that scientists call the lowest observed adverse effect level (LOAEL). This LOAEL occurs with the smallest amount of a substance found by experiment or observation to cause an adverse alteration in one's structure, function, development, or life span. It is at this level where things get fuzzy and unpredictable with organic toxins. At this lowest level, which might only be a microgram in the case of TCDD, each person, depending on their unique physical and genetic makeup, will react differently to the organic chemical(s) they are exposed to.

Take, for example, an adult who unknowingly swallows a tiny amount of strychnine, an alkaloid pesticide. Let's say they orally ingest 30 mg or

0.006 of a teaspoon of the chemical; they will die if not treated promptly. If we up the dosage to 120 mg (0.024 of a teaspoon), death is imminent. In fact, the first symptoms of strychnine injuries will appear within fifteen to sixty minutes after ingestion and quicker if the substance is inhaled or injected. Thus, with this kind of pesticide, there is no wiggle room.

Conversely, when you absorb a tiny amount of TCDD (about the size of a poppy seed), your internal cells are damaged or altered, but you don't die—yet. It may take many years or decades for the low-level slow-acting damage caused by exposures to morph into a diagnosable illness. However, after you have carried around the cellular damage for a decade or more, our government and chemical companies can effortlessly blame your illness on something other than their pollutants.

The bottom line is that there is considerably more wiggle room when the results are not as evident as they are with a strychnine pesticide. Perhaps this concept is best conveyed by a quote from the fifteenth-century physician Paracelsus, who said, "Poison is in everything, and nothing is without poison. The dosage makes it either a poison or a remedy."

Our federal laws do not require combined testing for the multitude of hazardous chemicals found in many common, widely used pesticides which can plausibly be present in our everyday environments. Regrettably, many of the most dangerous chemicals found in pesticides are not even the active ingredients but are those being categorized as inert. So how harmful can an inert substance really be? After all, the word *inert* implies that the material is inactive—just a chemical going along for the ride.

"Inert" Ingredients in Pesticides

Did you know that well over 1,600 active ingredients are used to create a multitude of different pesticides? However, the actual process of creating a pesticide consists of only two steps:

Step 1: Formulation of the chemicals

Step 2: The manufacturing process

Both steps involve the creation of planned reactions between chemicals in order to create an active ingredient for the particular pesticide being manufactured. The practices that occur within these two steps can potentially create toxic issues, harmful emissions, or hazardous waste materials, especially if the pesticide producer does not take proper safety precautions. Every pesticide is a specially designed combination of both active and inert ingredients. Active ingredients target and kill the pest or plant, while the inert components help the active chemicals work more efficiently and effectively, or so we've been told.

The fact is, these so-called inert components will not be tested as thoroughly (if they are tested at all) as the refined active ingredients might be. Plus, manufacturers are not required to disclose them on product labels, even though many inert ingredients may, on an individual level, be quite toxic. Thus, a chemical that's listed as an active ingredient in one pesticide creation can, conversely, be categorized as an inert ingredient in an entirely different pesticide product.

The ingredient classification of active or inert is contingent almost wholly on the manufacturer's designation. Consequently, in 1995, when the EPA evaluated a list of "inert" ingredients, it found that approximately 394 (or 16 percent) were or had been registered as active ingredients in other pesticide products.

In the real world—outside of laboratory applications—almost every inert ingredient is used to enhance the toxicity and application of the active ingredients. For that reason, "inert" erroneously implies that these additives and contaminants (e.g., dioxin, HCB, OSS-TMP, and other DLCs) do not have biological impacts or consequences. As a matter of fact, a 2006 study determined that inert additives actually make pesticides much more dangerous to humans than established by either our past or current safety testing standards.[8]

Regrettably, these last three chapters were just the small tip of the chemical iceberg. There are still many chemicals—both known and unknown—we veterans were exposed to; however, I think I made my point with just the few I noted. Besides, I'm pretty sure not everyone was having fun with this small portion of the chemistry that went into the Vietnam War. Even though a little complex, I hope these sections were enlightening and you gained some new insight into the number of toxic substances that veterans were really exposed to and the chemicals that our government didn't tell you about. Unluckily, not only were the actual chemicals problematic, but there were also severe complications with the spraying of those pesticides.

End Notes - Chapter 8

1. The United States Environmental Protection Agency - Title Superfund Record of Decision: Fort Dix Landfill September 1991

2. Cleaning up in New England - US Environmental Protection Agency 2017

3. Institute for International Urban Development - *Fort Devens* Case Study November 2008

4. EPA - *Fort Devens* Superfund site - Site Reuse Profile November 20061.

5. Veteran Administration - Exposure to Contaminated Drinking Water at Camp Lejeune

6. The United States Environmental Protection Agency – EPA Superfund Record of Decision: Hill Air Force Base 9/29/1998

7. Science Corps. - Veterans health - Health Hazards of Chemicals Commonly Used on Military Bases

8. Environmental Health Perspective - 2006 Dec; 114(12): 1803–1806 - Unidentified Inert Ingredients in Pesticides: Implications for Human and Environmental Health

Chapter 9
Vaporized Mist Drifts and Other Problems

Along with the inescapable toxicity of our military's herbicides and insecticides, severe problems also arose with their application. Theoretically, military aircraft were to fly at an altitude of 150 feet above the jungle floor at a rate of speed somewhere between 150 and 172 miles per hour. Once the proposed proper speed and altitude were achieved, the planes could begin spraying operations. In addition, the turbulence from the propellers was actually supposed to help distribute the atomized spray throughout the dense jungle foliage.[1]

All these best-case recommendations were—in theory at least—to minimize the drifting of the atomized herbicides or insecticides in the atmosphere. However, all the proposed preparations and procedures failed miserably in real-life applications; pesticide mist drifts were real problems for the military throughout the war.[2]

Numerous official reports chronicle unintended damage caused to rubber trees, banana plantations, other crops, and even ornamental shrubbery, but strangely, nothing was reported on the health impact on military personnel and civilians in the areas of those numerous vaporized mist-drift incidents. So while the military pesticide spraying programs hoped for a hypothetical best-case scenario, the real-world applications undoubtedly resulted in the drifting of hazardous chemicals.

One of the primary reasons for the drifting problems was that these so-called spray droplets were less than one hundred microns in size, which by most scientific standards would make them highly driftable. (Just as a point of reference, some everyday items that are around one hundred microns include pollen, coal dust, fine sand, a human sneeze, and dust mites.) So because of the minuteness of the atomized droplets, the resulting beads of chemicals would be more dependent on the natural movement of air currents than on gravity. They would just float on the breeze.[3]

Rough governmental estimates try to suggest that most of the pesticides sprayed in Vietnam reached the jungle; nonetheless, government scientists projected that 13 percent—more or less—was lost to drifting on natural wind currents. These guesstimates were based only on mission compliance with the correct speed, altitude, and weather requirements being exact. Then again, the probability the percentages were that low or remained constant are slim to none. Thus, the projected measurement could be several percentage points higher, which would substantially increase the amount of pesticide spray given to the wind and the distance the atomized droplets were capable of traveling.[4]

All the same, even if 13 percent of just the herbicide mixtures was given to the winds in the almost-weightless form, that would equate to well over 2.6 million gallons of concentrated wayward toxic chemicals— and that's assuming all the best-case mission requirements were met. While it is theoretically possible, the probability of only 13 percent of the herbicides and insecticides sprayed being lost to drifting is low, primarily because, in Vietnam, there were just no ideal days for aerial spraying. In point of fact, the mist drifts were incredible, even if we presume them ultraconservatively.[5]

Although vaporized mist drifts were persistent and serious problems for the US throughout the war, there was one incident that stole the spotlight. In November of 1967, Eugene Locke, deputy US ambassador to

Saigon, sent a memorandum that addressed the grave problems of drifting and the significant economic damage caused to various rubber and banana plantations by spraying earlier in the year. Keep in mind that Locke wrote this memo in 1967, after the war and spraying were both well under way:

> The range of drift of *vaporized* herbicide, however, *has not been scientifically established* at the present time. In recognition of this phenomenon and to minimize it, current procedures require that missions may be flown only during inversion conditions, i.e., when the temperature on the land and in the atmosphere produces downward currents of air. Estimates within the Mission of vaporized herbicide drift range from only negligible drift to distances of up to 10 kilometers and more.[6]

As Locke stated, the spray's traveling distances had not been scientifically established as late as 1967, so it isn't clear just how he determined that ten kilometers (6.21371 miles) was an adequate or even a safe distance. Even if we accept ten klicks as a ballpark figure, that would still mean any military personnel operating within a bare minimum of a six-mile radius or even more, depending on conditions, from defoliation or fly-swatter operations could find themselves, their water, and their food contaminated with vaporized mist drifts from whatever chemical agent(s) was being sprayed. The reality is that no one knew for sure how far the mist drifts would reach, because of all the unknown variables involved and the micro-size droplets.

Then in March of 1968, General A. R. Brownfield (the US Army's chief of staff) sent a memo to all senior US advisers in the four tactical war zones of Vietnam regarding vaporized mist drifts and their continuing problems. General Brownfield ordered the following in his memo:

> Helicopter spray operations will not be conducted when ground temperatures are greater than *85°* Fahrenheit and wind speed in excess of *10 mph*.[7] [In reality, 85° was below the average daytime temperatures in Cam Ranh between March and October.]

Regrettably, vaporized mist drifts could have contained any pesticide from Agent Orange contaminated with 2,3,7,8-TCDD to Agent White laced with HCB to malathion adulterated with injurious DLC, and we would have no way of knowing because the origins of these drifts were miles away. While we can give mist drift distances a good guesstimate, accuracy is another story. What is quite evident and can be proven is that harmful pesticides were, without a doubt, atomized and could wander long distances, almost weightlessly, on natural currents of air.

In addition to pesticide mist drifts, we had to contend with "dust drifts" from bombings and napalm as well as "smoke drifts" from burning dung, trash, and defoliated areas that had previously been sprayed with an herbicide, such as Operation Sherwood Forest or Project Pink Rose. (Sometimes I wonder if our military leaders stayed awake at night dreaming up all these fancy-sounding operational names.)

Whatever the case may be, Operation Sherwood Forest, according to now-declassified documents, was an experimental bombing mission conducted on Boi Loi Forest, about twenty-five miles west of Saigon, in March of 1965. While this initial test operation was considered unsuccessful, it continued to stimulate interest in killing and drying out vast areas of jungle to the point where the foliage would be combustible. From the failed test at Boi Loi Forest, the military cooked up their next ingenious strategy.

Once again, according to documents declassified in 1990, it seems that even though the Sherwood Forest mission was a failure, it did induce sufficient "fire" to encourage the military hierarchy—the ones who wanted the ability to destroy vast areas of the jungle by burning them—to continue their clandestine endeavors. Their resolve finally culminated in two test operations, which were conducted in March of 1966 near Pleiku.

The Pink Rose tests consisted of fifteen B-52 bombers dropping 750-pound M35 incendiary cluster bombs on an area of jungle that had been sprayed with an herbicide—Agent Orange or its equivalent. According to the declassified memo, "The results were inconclusive, but sufficient fire did develop to indicate that this technique might be operationally functional."[8]

While I'm not an expert in the field, I'm pretty sure that 750 pounds of combustible material, such as a bomb with a white phosphorous fuse igniting hundreds of pounds of napalm, will burn up everything in its path, short of water. So it's hard to visualize what that account is trying to establish. However, what the report doesn't say is also very critical. In reality, what our government and chemical companies knew (or should have known) was that burning areas sprayed with an herbicide or any substance containing dioxin compounds would significantly increase their toxicity. Thus, not only did the military hierarchy continue to introduce hazardous chemicals into the war, they raised their toxicity level by over 25 percent just by burning them.[9]

The Vietnam War is, without a doubt, a war that will not end. It's a war that continues to attack, injure, and claim soldiers and their progeny decades after the last shots were fired. It was a war of multicolored herbicides, malathion, and a host of very hazardous chemicals, all of which were just as deadly as the adversary we fought there—only organic chemical pesticides were a whole lot slower, far more deliberate, and considerably more treacherous than our human adversaries were.

Problems with Our Food and Water

Over the years, I've read through a lot of Operational Lessons Learned military command reports and found that they contain excellent information about the Vietnam War. For instance, in the following excerpt from a report written in 1968 by US Army Support Command Headquarters, Cam Ranh, they discuss pigs and milk production in-country. You got it right—pigs and milk. I'm not sure what the lessons were, but you just can't make this stuff up:

> Two projects are now in the final stages of planning that will form the basis for new and self-sustaining industries in the greater Cam Ranh City area. The Pork Production Program, as a civic action, will provide edible garbage for approximately 4,500 pigs in this area, alleviating a critical feed shortage and increasing pork production, ultimately, to 10,000 head.

> As a long-range program, the second project will create a truck farm to provide Cam Ranh Bay with locally grown fresh vegetables. Starting as a civic action, it should result in a major means of livelihood for much of the local population.[10]

The commercial milk plant at Qui Nhon became operational on 15 February 1968. This plant provides a highly acceptable filled milk product, ice cream, and cottage cheese.[11]

I could go on and on, but why belabor the point? Unless the US military covered all our potable water and imported all our fresh food supplies, they would have been procured from local Vietnamese sources and also very problematic. The fact is that our water for cooking, drinking, and showers was very probably polluted with any number of different chemicals. In addition, when we were finally able to acquire food items such as eggs, milk, butter, ice cream, bread, fish and meat, fruits and vegetables, they were undoubtedly contaminated with any number of harmful chemicals. Hence, all our locally procured dairy and fresh-food items, as well as our water supply, were great conduits for the entry of any of the numerous harmful pesticides and organic chemicals used and sprayed in South Vietnam, some of which were noted in study reports compiled about the "Blue Water Navy."[12]

Blue Water Navy

Once again, while I want to keep my focus on in-country veterans, it would be a slipshod investigation if I didn't include what I discovered about US Navy personnel. While many feel that anyone who didn't set foot on the soil of Vietnam shouldn't receive the same benefits as boots-on-the-ground personnel, which may not be the case. The so-called Blue Water Navy (BWN) were naval personnel who were stationed on ships in the South China Sea or the Pacific Ocean and who assisted the US during the war, but they were well off the coasts of Vietnam. Thus, they were never sprayed with the same chemicals that in-country veterans were, and they didn't have the same stresses or living conditions that we did. Nevertheless, they were exposed to some of the same very toxic chemicals.

You are probably asking, as I did, "How is that possible if they never set foot in Vietnam?" Well, during my research, I also found several reports and investigative studies on the BWN. One such account was assembled by the University of Queensland in Australia. According to their evidence, Australian BWN veterans were actually exposed to some of the same pesticides that in-country veterans were, only by a very different delivery system. The Australian investigation revealed that the water desalinization units of both Australian and American ships had a contrary effect. It seems that while the desalinization units were removing salts and minerals from seawater to produce drinking water, they concentrated harmful organic chemicals in the process.[13]

The study went on to explain that it was discovered that the herbicide mixtures being sprayed on the mainland of Vietnam—to which either kerosene or JP-4 fuel was added—would float on or near the surface of any water source. As a result, the harmful chemicals contained in runoff or sprayed on the ocean directly would be carried many miles out to sea by the wind and ocean currents. It was there that the contaminated water would be taken in by BWN ships and the dioxin and other organic chemicals concentrated by the desalination units aboard those vessels. So even without setting foot in Vietnam, our BWN was contaminated by many of the same oil-based pesticides and organic chemicals that we were.

Now to be fair, our navy brothers and sisters were also exposed to many different toxic materials that we weren't, substances such as asbestos, cleaning solvents, and other chemicals needed to operate and maintain their ships. However, that's a whole other research endeavor, and I do want to stay focused on boots-on-the-ground veteran issues. So we will leave the BWN here and move on, and it's "Fair winds and following seas and long may your big jib draw" to all my navy readers.

We Few, We Unlucky Few, We Band of Brothers

Now let me get back on track. Thus far, I have made many references to in-country veterans as a group being different from other veterans of the Vietnam War era, so who are we really and why are we distinctive? We are, beyond a shadow of a doubt, a class apart. We are unique in the history and annals of war, toxic chemical exposure, and harsh jungle environments. "How so?" you might ask. "After all, Vietnam wasn't our country's first jungle war."

First and foremost, we served in—hands down—the most unpopular and widely opposed war in the history of our nation.

Second, it was a war that our government, from the very beginning, knew that we couldn't win—as illustrated by the Pentagon Papers.

Third, it was a war predominately intended to prop up the failing regime of South Vietnam, ostensibly a civil war.

Fourth, it was the first war in which we were subject to the hit-and-run terrorist tactics of the insurgent Viet Cong from within a civilian population. We didn't know who the enemy was or where they might strike.

Fifth, we were knowingly exposed to numerous physically damaging and cell-altering organic chemical substances by our government and military leaders.

I could continue pointing out more reasons for our distinctiveness, but I think you get the picture I'm trying to paint. My last point, however, is without question the most damning of them all, primarily because our government leaders would deny the harmful effects of all the multicolored sunset pesticides they allowed to be sprayed in Vietnam.

While it's appropriate—due in large part to enormous pressure from veterans' groups, Congress, and members of the public—that several illnesses and disorders have finally been presumed by our government to be associated with our service in Vietnam, still, these few acknowledged illnesses and disorders (diabetes, ischemic heart disease, Parkinson's, a few different types of cancer, and Lou Gehrig's disease) are nowhere close to being a comprehensive list.

So you might be asking yourself, as I did, "Exactly how can the DVA realistically or legitimately determine the presumptive probabilities of the illnesses and cancers being produced by the pesticides and conditions to which veterans were exposed without actually studying them?"

If you are like me, then that question is a bit disconcerting, and the short answer is: They can't. Nonetheless, our government and DVA are doing precisely that even in spite of the fact that it is virtually impossible for the DVA to presume a statistical probability of less than 50 percent without first researching all the entangled, highly toxic organic chemicals used in Vietnam and basing that examination on actual veterans who served in Vietnam.

End Notes - Chapter 9

1. The National Academies of Science - Veterans and Agent Orange: Health Effects of Herbicides Used in Vietnam – Chapter 3 - The U.S. Military and the Herbicide Program in Vietnam.

2. United States – Vietnam Scientific Conference on Human Health and Environmental Effects of Agent Orange/Dioxins – The history of Agent Orange use in Vietnam.

3. Institute of Medicine (US) Committee on Blue Water Navy Vietnam Veterans and Agent Orange Exposure. National Academies Press; 2011 chapter 4

4. Ibid

5. Blue Water Navy Vietnam Veterans and Agent Orange Exposure - Chapter 4 – Fate and Transport of Herbicides used in Vietnam

6. United States – Vietnam Scientific Conference on Human Health and Environmental Effects of Agent Orange/Dioxins – The history of Agent Orange use in Vietnam

7. Ibid

8. The Pentagon's Brain: An Uncensored History of DARPA, America's Top-Secret Military Research Agency by Annie Jacobsen

9. United States – Vietnam Scientific Conference on Human Health and Environmental Effects of Agent Orange/Dioxins – The history of Agent Orange use in Vietnam

10. National Archives - Operational Reports - Lessons Learned, Headquarters, US Army Support Command Cam Ranh, Period Ending 30 April 1968 – Page 3

11. National Archives - Operational Report - Lessons Learned, Headquarters, 1st Logistical Command, dated 21 August 1968

12. Institute of Medicine (US) Committee on Blue Water Navy Vietnam Veterans and Agent Orange Exposure. National Academies Press; 2011 and A Re-Analysis of Blue Water Navy Veterans and Agent Orange Exposure

13. Ibid

Chapter 10
The Biology of Vietnam

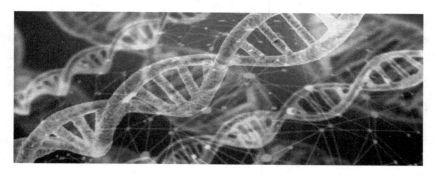

We've taken a long hard look at the chemistry of Vietnam and how the pesticides were delivered. Now we must examine what the biological influences of those organic chemicals were. Once again I will try to make my explanations as uncomplicated as is realistically possible. The problem is where to start. So, throwing a dart at my investigation work board, we will start with how all the organic substances affected the cellular structures of veterans, or what scientist would call the environmental epigenetics of Vietnam.

Now you might be asking yourselves, "Just what the hell is environmental epigenetics, and what does it have to do with the Vietnam War?" In the purest sense, it's the investigation of changes in gene expression that do not involve changes in the DNA sequence. While environmental epigenetics sounds and is indeed challenging and complicated in real-world applications, it can be explained in one short paragraph:

> Environmental epigenetics involves the ability of environmental factors—such as nutrition, toxicants, and stress—to alter epigenetic programming. Thus, epigenetics provides a molecular mechanism by which environmental factors can influence disease etiology. The role of epigenetics in disease etiology has been shown for cancer and a number of other diseases.[1]

The vast array of toxic substances veterans were exposed to in Vietnam were all capable of altering our epigenetic encoding. As resilient as the human body is, it can only compensate for coding damage and cell-altering substances for a limited period and even less under stressful conditions, poor health, or as we age. Eventually, our bodies will reach a saturation point where the damage done by organic chemicals to our cells will become overwhelming.

To use an exaggerated analogy, our contact with and exposure to the different pesticides used in Vietnam can be likened to holding an empty highball glass in the air. Your body and muscles have no problem picking up the light glass and holding it out. Now, after you have the glass up, a little water is added to your glass, and then a little more. As time goes by, the glass continues to be slowly filled with water. However, the more water that is transferred into the glass and the longer you have to hold it up, the greater the burden on your muscles. After you have been holding on to the glass for a couple of hours, the water starts to overflow. After a day, it becomes heavy. As time goes on, that same simple and almost insignificant glass of water becomes a substantial problem until it reaches a point where the body can no longer bear the burden; physical systems start to break down, and the glass falls. Depending on the unique, innate attributes of each person, this process may take a week, a year, or perhaps even a decade. A similar comparison can be made regarding exposures to all the cell-damaging compounds contained in herbicides and insecticides used during the war.

The impact on the human body is minimal at first, depending on the dosage (LOAEL) and the criteria discussed in the next section. Nevertheless, the long-term consequences of contact with and internal storage of cell-damaging environmental chemicals are analogous to my little water glass scenario. As a result, we cannot compare toxic chemical exposures to the visible physical trauma of being hit by a truck or shot with a bullet. The consequences of cellular damage created by pesticides are far more subtle and sinister—although ultimately just as injurious and deadly—only a whole lot slower. Just like the little glass of water, as more toxicologically destructive toxins are added to our cellular structures—the longer they have to be carried by biological systems—the more significant the strain placed on our whole body. Eventually, our cellular structures reach an overload point where they can no longer compensate for the impairment and damage. It is at this point that the affected system(s)

110

begin to break down, and an actual diagnosable illness or disorder becomes apparent.

The harmful chemicals we were exposed to in Vietnam produced our initial unquantifiable trauma—clandestine, unfelt, cellular, hormonal, and genetic damage and changes. However, it was our own unique individual strengths and weaknesses, health, and genetics—plus the additional processes of time and aging—that determined precisely when and what cancers, illnesses, or disorders would ultimately become apparent and exactly how long it would take for that to happen. It might take years or even decades, depending on the unique strengths and weaknesses of each individual as well as on the following exposure criteria.

Pesticide Exposure Criteria

The detrimental health impacts and cellular changes produced within our immune, endocrine, respiratory, neurological, and cardiovascular systems by the chemicals we veterans were exposed to were dependent on numerous complex, interwoven factors and events. Therefore, calculating them or picking just one is nearly impossible. The following are some primary exposure parameters and criteria developed by the National Academy of Medicine (NAM, formerly the Institute of Medicine) and other scientific organizations:

1. The total amount of dioxins and other chemicals absorbed or assimilated

2. The period of time over which the substances were absorbed, plus internal exposure storage factors

3. The method of entry (air, water, food, skin)

4. The genetic and cellular predisposition (strengths or weaknesses) of the individual exposed

5. The age, health status, and stressors of the individual at the time of exposure/contact

6. The synergistic interaction of other pesticides or chemicals to which exposed[2]

These six points are the cold, hard scientific standards for defining and determining pesticide exposures. They are measurable factors, which are

111

essential in finding out how all the various pesticides affected veterans. Thus far, almost none of this statistical data has been available because of sloppy record keeping, document classification, and stonewalling. While it is an established fact that veterans were exposed to many dangerous, injurious pesticides during the war, the health impacts of those chemicals have not been studied in earnest, especially on a genetic/cellular level.

Our Cellular Coding Systems

Biologically speaking, each and every cell in our uniquely designed bodies contains somewhere in the area of thirty thousand genes. Most of these genes are responsible for inherited traits. However, there are some four thousand—more or less—genes that are active at any given moment, controlling and guiding specific cellular functions. Take, for example, that right now in any healthy human liver, roughly two hundred genes are turned on to guide the function of each liver cell. At the same time, in our immune system, these same two hundred "liver genes" are turned off. But, on a different part of the DNA structure, another set of approximately two hundred genes are turned on to control the characteristics of the individual immune-system cell.

These on-and-off genes control such things as how individual cells form connections with other cells. They regulate how each cell produces essential compounds to fight viruses and cancer. They also determine how each cell distinguishes between foreign cells and friendly self-cells. In other words, our cellular structure has the complex and essential job of maintaining our health every second of every day we are alive.[3]

In addition to our double-helix DNA configurations, we also have single-strand ribonucleic acid (RNA) molecules in our cellular makeup. These two nucleic acids work in conjunction with each other in a fashion similar to how inorganic computers and their programming operate. In an analogous manner, our DNA duplicates and stores genetic data (lines of code): an outline and instructions on how to produce every minute part, cell, hormone, and process within our bodies. Equally important is our cellular RNA, which translates the genetic info in our DNA to a format our bodies can understand and use to build proteins. In the simplest terms possible, RNA acts as a messenger to translate information and carry it from our DNA to control the synthesis of proteins by our cells.

This intricate codifying system governs and sustains the human body from conception to death. If any of the elaborate links or portions of the body's messaging system are broken, contaminated, or damaged by exposure to toxins, then the impaired encryptions will be used when the body produces new cellular structures to interchange with the old ones. As a result, the replacement cells will also contain the defective coding. In fact, you might even characterize pesticide contamination as throwing a monkey wrench—albeit a tiny one—into the cellular "machinery" of the human body. Likewise, you can realistically compare being exposed to dioxin or any organic chemical to being exposed to ionizing radiation.

Again just as an illustration, let us suppose a shipment of nuclear waste is to be transported secretly, and you've been hired to drive the truck carrying the materials from the US East Coast to the West Coast and back again. During the trip, you're unwittingly exposed to ionizing radiation leaking from the substandard containers you are transporting. Sometime later, depending on the dosage and time of exposure, you are tired and lose your appetite. You start to have headaches and begin to vomit. You go to your family doctor, who tells you that you have a stomach virus. He gives you a couple of prescription medications, and coincidently, you start to feel a little better. Months pass by, and now you have adapted to being tired and feeling sick all the time. Your conditions may even start to feel reasonable to you.

As time goes on, you stop wondering and just accept your conditions. Your body continues to bear the internal damage caused by the ionizing radiation. Shortly afterward, you're diagnosed with a serious illness, such as stomach cancer or leukemia. Coincidently, at the same time, you also discover information about your previous exposure to ionizing radiation. So now, equipped with the knowledge of your exposure, all your health problems start to make sense, and you begin to piece them together.

It was your exposure to ionizing radiation that caused your symptoms and made you feel sick all the time. However, neither you nor your doctor could have linked the illness and ionizing radiation together unless it was known beforehand that you had been exposed to that type of emissions. The very same can be said of being exposed to pesticides—both herbicides and insecticides. The initial evidence of exposure damage (cellular and genetic) is invisible and can be carried for years or even decades before it builds up to a point where it can be recognized as an illness or disorder.

113

Unfortunately, damage to our cellular coding system wasn't the only biological problem we brought back from Vietnam. The array of systemically damaging organic chemicals also had negative impacts on our gene expression, cellular receptors, and even our hormonal systems.

End Notes - Chapter 10

1. The National Academies of Science - Veterans and Agent Orange: Update 2012 (2014) Epigenetics is the study of changes in organisms caused by modification of gene expression rather than alteration of the genetic code itself.

2. The National Academies of Science - Veterans and Agent Orange: Update 2012 (2014) Chapter 4 - Information Related to Biologic Plausibility pages 80–115

3. Experimental Gerontology - The Human Aging Process: Gene Loss as the Primary Causes

Chapter 11
Systemic Toxicants

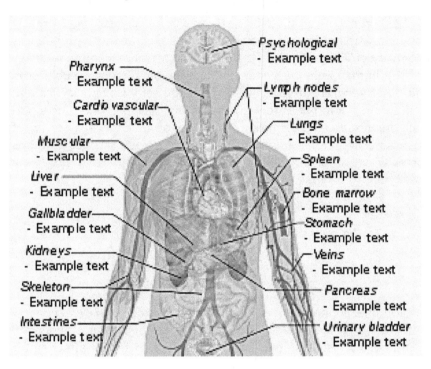

I have talked a lot about the systemic toxicity of the organic chemicals used in Vietnam, but what exactly does it mean? A systemic toxin such as dioxin is summed up in one word: xenobiotic. Now wasn't that simple?

Actually, "xenobiotic" is just the short scientific way of saying "a foreign substance taken into the human body that will affect all its organs and biological systems to varying degrees, with the major effects being manifested in one or two systems." I think I like the word *xenobiotic* better. With that said, let me prepare you for this chapter, which may read and sound to some like a tedious course in biology. Regrettably, I do not know of any simple way of explaining the actual impact dioxin, DLC, and all the other contaminants had on veterans and their cellular structures.

Virtually every medical study I have researched indicated that dioxins will directly or indirectly influence all the biological systems of the human body, but they have a particular fondness for cells found in the human immune and endocrine systems.[1] Tragically, toxic chemical injury to either of these two critical physical systems will be devastating to the entire body. So working from that evidence, let's begin with the influences the chemistry of Vietnam had on the biology of the human immune system.

The human immune system is an assembly of cells, soluble proteins, and other general defenses, which have the astonishingly intricate ability to accurately communicate with each other. Our immune function, when well organized, is the body's main line of defense against infectious organisms and other invading substances. When it does not operate correctly, we have a problem. Overall, immune system illnesses can be placed into four broad categories:

1. Immune suppression

2. Autoimmunity

3. Reaction to foreign substances (i.e., allergies)

4. Inappropriate inflammation responses

Immune suppression is generally associated with infections or an increased risk of cancer. However, autoimmune, allergic, and inflammatory disorders can be manifested as diseases affecting almost any type of tissue. Both immune and autoimmune illnesses are often challenging to identify, so they may or may not always be medically classified as immune system disorders.[2]

At present, medical science is still trying to understand better why the cellular alterations and inflammatory processes that fuel our intricate cancers and immune disorders occur in the first place. Nevertheless, low-level environmental exposure to pesticides is right at the top of the scientific list of primary causes. In fact, a scientific review of real-world pesticides in 2002 found that malathion, lindane, and aminocarb can cause adverse autoimmune and allergic reactions in sheep and mice.[3]

Additionally, it is essential to understand that if exposure to dioxins— even temporarily—raises the body's dioxin levels when our immune system defenses are initiated, this activation may increase the risk of adverse harmful impacts to our cells. This phenomenon may occur even though blood test results show that the total level of dioxin is still very low or even almost nonexistent.[4]

In other words, a single microscopic dose of dioxin—less than a grain of table salt—ingested or absorbed at the wrong time is capable of damaging the immune system's normal ability to protect our bodies and keep us healthy.

Take, for example, ulcerative colitis, an immune system illness with which I am personally acquainted. I can tell you from experience that UC is an unpleasant chronic immune system disorder that involves an abnormal response of the immune structure that adversely affects the large intestine. Under normal conditions, the cells and proteins that make up our immune system protect us from infections and foreign cells or materials. However, the immune systems of people with UC mistakenly identify the cells of the large intestine as foreign or invading substances and do not identify them as a part of the body. When this occurs, the immune system will send white blood cells to the lining of the intestines to fight the misidentified invaders. Thus, the attacking white blood cells produce chronic inflammation and, over time, ulcerations.

Many studies have been conducted on the impact of systemic toxins such as dioxin on the human immune system. According to many of these studies, quite deceptively, dioxin is a two-edged sword. On the one hand, dioxin will suppress our immune systems, resulting in the inability to fight mutant cancer cells or infectious maladies. On the other hand, it can upregulate (excite) our immune systems to produce hypersensitivity, autoimmunity, and allergies.[5]

Consequently, in an immune system that is severely suppressed by pesticide exposure, the compromised cells are not recognized as damaged or foreign and are allowed to multiply freely, thereby replacing healthy tissue with mutagenic cells. Conversely, depending on the already-noted exposure factors, toxic pesticides, especially dioxin-TCDD, can also excite the immune system. This upregulation will cause the immune system to respond in such a way that the body will attack itself by increasing its inflammatory processes in an attempt to destroy what it perceives as foreign material. Thus, the likelihood is very high that pesticide exposures in Vietnam are the primary source of countless cancers, illnesses, and inflammatory disorders, with UC being only one of many.[6]

Biological Receptors and Gene Expression

The medical science behind the chemistry and biology of the Vietnam War is absolutely astonishing while being horrifying at the same time. Similarly, the cellular structure of the human body is nothing short of miraculous. The cells in our bodies react to all the different changes in our environment and are able to receive and process various stimuli that originate internally as well as externally. Hence, individual cells will receive a wide range of inputs at the same time, and they have to correctly integrate all the information accumulated for our bodies to function normally. But they aren't just targets of incoming data. They are also sending out messages and coordinating with other cells.

For the most part, incoming and outgoing cell signals are biochemical in nature. In human beings, hormones and neurotransmitters are just two of the many types of chemical communicators our cells use. These biochemical substances can exert their effects locally, or they can travel throughout the whole body. Some cells also respond to mechanical stimuli. For example, sensory cells in the skin react to the pressure of touch, while similar cells in the ear respond to sound waves. Additionally, our bodies have dedicated cells, such as those in the cardiovascular system, which detect changes in blood pressure and then use that information to maintain a consistent cardiac load and general homeostasis.

Each cell of the human body comprises many different substances, with one of those constituents being protein receptor molecules. Each different receptor is specific to a particular ligand. For instance, insulin receptors bind with insulin, acetylcholine receptors bind to the neurotransmitter acetylcholine, and so on. In point of fact, there are numerous receptor types throughout the cellular structure of our bodies.

According to epidemiological and medical research, several of the chemicals used in Vietnam will have wide-ranging adverse impacts on our cellular and genetic structures; however, investigators have also found that one of the most critical cellular consequences of pesticide exposure involves the aryl hydrocarbon receptor (AhR).

Yep, you guessed it: an AhR is a protein molecule in our bodies that is predetermined by the AhR gene. These particular receptors are heavily involved in the process of transcribing DNA into RNA, which in turn regulates gene expression. In addition, these receptors not only control our metabolism enzymes, but also have roles in regulating our immune system, stem cell maintenance, and cellular diversity.

Numerous medical and scientific studies have shown that dioxin and DLC will directly or indirectly affect many types of biochemical receptors. Furthermore, they will also adversely influence and disrupt the equilibrium of almost all the biological systems in the human body. As a result, dioxin exposure crosses through a number of scientific areas of expertise, such as toxicology, endocrinology, and molecular biology.[7]

Explaining how dioxins work in humans without a dreary toxicology course is a tough job because of all the complexities involved. So in a nutshell, our individual cells—as a part of their normal functioning—have thousands of surface biochemical receptors that are comparable to a loading dock system. Our cellular docking structure functions by allowing beneficial chemical compounds access into our cells. This chemical connection occurs on the cell's surface and can best be described as a lock-and-key arrangement. Once a cellular receptor is paired with its chemical key, either real or mimicked, it can then influence the DNA expression within that cell. This docking structure helps our bodies organize beneficial chemicals to enable healthy systemic functioning.

However, in the case of organic chemical contaminants, once the toxin binds to the AhR, the receptor and the mimicked key move into the cell's center—the area where genes are activated. The body will then produce enzymes to try to break the connection and get rid of the unwanted chemical hitchhikers. Unfortunately, numerous studies have established that after these substances, such as dioxins, bind to the AhR, dislodging them is extremely difficult. In fact, among the various identified organic toxic substances, TCDD and dioxin-like compounds have some of the tightest binding dynamics with human AhR.[8]

Thus, trying to pin down precise individual adverse health problems and associations is extremely difficult, if not impossible, because of the exceptionally complicated biochemistry involved. What's more, the cascading torrent of biological influences taking place within the individual cells of our immune, cardiovascular, respiratory, neurological, and endocrine systems also adds to the overall destructive impacts and complexities of these organic hitchhiking chemicals. Nevertheless, binding to the AhR is the first stage in a cascading sequence of biochemical and cellular events that ultimately leads to adverse health responses.[9]

Sooner or later, the biological systems of almost every veteran who served in Vietnam will be negatively affected by the organic contaminants to which they were exposed during the war.

Endocrine System

Today in 2019, there is no longer any question that the pesticides used in Vietnam were endocrine system disruptors: it is a fact. However, identifying the mechanisms by which they affected the human endocrine system is still highly problematic because of the extent of overlapping or cascading influences that take place within the human body. As a result, scientists are still having trouble pinpointing exactly how these pesticides disrupt hormone interactions, but they concur that the AhR and insulin receptors, which control blood sugar levels, are undoubtedly involved in that multifaceted process.[10]

Initially, pesticide-induced endocrine system damage will start with invisible system-wide cellular and genetic alterations. These unwelcome changes will in turn produce biological injury to insulin receptors, as well as damage to the way carbohydrates and fats are metabolized. Over time, these hidden wounds will cause our bodies to become more and more resistant to the insulin being produced. The increasing cellular insulin opposition, in turn, will create progressively higher fasting blood glucose levels (FBGL), or what my doctor would term hyperglycemia or impaired glucose intolerance (IGT), both defined as FBGL over 100 milligrams per deciliter (mg/dl) but under 126 mg/dl.

In due course, hyperglycemia will give rise to IGT, which in turn creates greater cellular insulin resistance, which then produces more significant damage to the endocrine system as well as overlapping damage to the circulatory and neurological systems. The whole process is like a snowball rolling down a hill, picking up snow as it goes, getting bigger and wider as the damage moves and expands. The end result is that long-term low-level exposure to pesticides will almost certainly increase the risk of developing type 2 diabetes, which is medically defined as an FBGL over 126 mg/dl two or more times, as well as other endocrine system disorders.[11]

In summary, the pesticides veterans were exposed to in Vietnam did produce invisible cellular changes that will increase the probability of their developing type 2 diabetes, period.

Type 2 Diabetes

Well, what do you know? Unluckily, type 2 diabetes is another illness with which I'm personally acquainted. Type 2 diabetes initiated by exposure to the harmful substances used in South Vietnam will start years—if not decades—before the first diagnosable symptoms ever appear. It begins with subtle and unseen yet very real cellular changes in the endocrine system as well as the immune, circulatory, respiratory, and neurological systems. So even before a diagnosis of type 2 diabetes, the pesticide-induced cellular damage is hard at work, systemically affecting the liver, heart, blood vessels, nerves, eyes, and kidneys, years in advance of FBGL reaching above the 126 mg/dl threshold. The actual diagnosis of type 2 diabetes is an exceptionally lengthy, far-reaching, and uncomfortable voyage. It is a journey with many negative health consequences and impacts along the way.

By way of example, the beginning of type 2 diabetes can be compared with starting a journey—albeit not a pleasant one. You begin your trip in the area my doctor would call prediabetes, or stage 1 diabetes, which is defined as an FBGL just beginning to reach above 100 mg/dl. Once you start out on your journey, you slowly progress, up and down, left and right, toward a higher and higher FBGL. At the same time, there is serious systemic damage being done to all the other biological systems of your body along the way.

The road between stage 1 diabetes and type 2 diabetes is a long, winding, broad, and bumpy one. All things considered, we do not go from pesticide exposure to type 2 diabetes overnight. It is an extremely lengthy process with many complex, confounding variables and a great deal of systemic damage along the way.

Aging of the Veteran Population

One inevitable fact of life is that we are all getting older whether we like it or not. Take me as an example. My brain still thinks it is a twenty-year-old kid, but when I try to do some of the exercises I did in my younger days, my body will say, "Whoa, Bubba, you're no spring chicken anymore." It simply refuses to cooperate with my mind. So, too, as I age, my cells become less efficient, and my body is less able to carry out its normal functions. For example, my muscles have lost much of their strength over the years, and my hearing and vision have become weaker. My physical reflexes have slowed down, the capacity of my lungs has decreased, and even my heart has been affected. Aging has also weakened my immune system, making it less able to fight infection and disease. Aging is definitely not for the faint of heart.

Unfortunately, there is no single concept to account for all the aspects of how we age; most medical theories can be divided into two schools of thought. The first is the belief that we are genetically programmed to age and die. Even with this encoded theory of aging, environmental agents such as chemical toxins can speed the process of aging along. This first concept is pretty straightforward. The end result is that our bodies and cells are predestined to die, and organic chemical pollutants just accelerate that process.[13]

The second major concept is that the natural wear and tear of life are causing us to age and eventually die. With the wear-and tear-theory, it's not the age but rather the mileage that gets us. This model suggests that cells and genes are negatively altered by random modifications in their structure that accumulate over time, gradually leading to diseases and, ultimately, death. Needless to say, environmental organic toxins add significantly to adverse physical changes in either theory.[14]

Many scientific publications agree that this second concept of aging is a little more accurate. Still, how exactly does the process of aging correlate with chemical exposures in Vietnam? Let me start by developing a thought-provoking but abstract study in which I can establish that two variables will consistently correlate with each other, but a causal relationship can be found with a third variable. A simple illustration of this relationship is a theoretical study conducted on the correlation of reading skill and shoe size across the whole population of the United States.

If someone, in fact, did perform such a study, they would logically find that individuals with larger shoe sizes have a better reading ability. However, this does not prove that large feet produce good reading skills. On the contrary, it is the fact that young children have small feet and have not yet been taught to read that produces the conclusion. In this little made-up scenario, the two variables—shoe size and reading skills—are more accurately linked with a third important factor: age.

Age, or the process of aging, is known as a variable factor in this little hypothetical study, and it must be accounted for. Age affects both reading ability and shoe size quite directly. Thus, aging can also be the actual cause of a correlation, as in this hypothetical case. In theory, a

122

relationship between reading ability and shoe size must also include age as a factor, not as a confounding influence. It is the same with the pesticides we aging in-country veterans were exposed to. The initial cellular and genetic damage and alterations produced by unprotected exposure to all the tactical military chemicals must also be investigated in the context of getting older.

As noted earlier, herbicide and insecticide exposure is not like being shot or hit by a truck, where the impacts are immediate, visible, and dramatic. Pesticide exposures are far more complex, more like the aforementioned ionizing radiation exposure analogy. The cellular and genetic alterations and dose-dependent damage associated with ionizing radiation are initially unseen and unfelt, but we know both scientifically and biologically that they occur.

Unfortunately, in-country veterans have very limited information on all the substances contained in the pesticides we were exposed to and even less on how our health was adversely affected by the chemistry and biology of the Vietnam War. So how do we find out what the facts and the truth are regarding the health consequences of the pesticides we veterans were exposed to in Vietnam?

End Notes - Chapter 11

1. Dioxins and the Immune System: Mechanisms of Interference - Allergy Immunol 1994;104:126–130

2. The National Academies of Science - *Veterans and Agent Orange: Update* - Page 272

3. *Research links Pesticides to Autoimmune Disease* by Toxic Free NC volunteer Amy Freitag

4. Rachel's Environment & Health Weekly # 414 - *Potent Immune System Poison: Dioxin* - November 3, 1994

5. Ibid.

6. The Board of Veterans' Appeals Decision search

7. Department of Molecular Pharmacology, Stanford University School of Medicine, Stanford, CA - *The aromatic hydrocarbon receptor, dioxin action, and endocrine homeostasis*

8. Journal of Biology - Beyond toxicity: aryl hydrocarbon receptor-mediated functions in the immune system - Brigitta Stockinger - 17 August 2009

9. Environmental Protection Agency - Recommended Toxicity Equivalence Factors (TEFs) for Human Health Risk Assessments of 2,3,7,8-Tetrachlorodibenzo-p-dioxin and Dioxin-Like Compounds (Recommendation Section - Page 12).

10. Diabetes and Environment.org - Diabetes and the Environment:Dioxin

11. Ibid.

12. Science Clarified - *Aging and death*

13. Ibid.

Chapter 12
Truth, Facts, and Probabilities

$$P(A|B) = \frac{P(A)\,P(B|A)}{P(B)}$$

Bayesian Probability Equation

Facts and *truth* are two simple words that have long been the playground of philosophers and theologians. While many academics have sought to understand and define the difference between facts and truth, for most of us today, the truth is merely the opposite of a lie. So, like me, you might be tempted to ask, "But is what we call truth just a simple matter of true and false? What's the relationship between 'the facts' and 'the truth?' Are they both the same, or are there differences between them?" The most uncomplicated answer I can give you is "Yes, there are contrasts."

The primary distinction between the two words is pretty straightforward. Facts are bits of information comprising statistical and scientific data. In comparison, *truth* refers to the validity of the collected facts and depends almost entirely on an individual's perspective and life experiences. While the word *fact* is sometimes used interchangeably with *truth*, there are several differences between them. But to a large extent, these variances are less about terminology and more about individual viewpoints.

The two words can be roughly equated to the difference between a pile of wood (facts) and a house (an arrangement of facts that would equal truth). An isolated fact is like a stray piece of a puzzle; it's a fragment of information. Truth, alternatively, is all about putting facts together and giving meaning to them, or to be more concise, discerning the truth is a matter of interpreting the facts.

Take for example what happens in every courtroom in the US during any civil or criminal trial. In the courtroom, the same independent facts are available to both sides. Each attorney gets a chance to build their own pattern of "the facts" or, if you prefer, evidence. While both sides have the same evidence to work with, there is only going to be one victor, depending on how the judge or jury understands the relationship of the

125

facts to the case presented. Individual facts can be tested or checked. We know water freezes at 32 degrees Fahrenheit; the speed of light in a vacuum is 186,282 miles per second; the organic chemicals contained in the pesticides used in Vietnam were injurious to humans. All facts.

Conversely, the truth is an opinion or preference. For example, the fact is that the temperature in my living room is seventy-four degrees. The truth is I'm comfortable. The truth is my wife, Georgia, is cold. The truth is my son is warm. One fact, three different realities. Who's telling "the truth"? In all the cases, the seventy-four-degree fact remains constant. The same is true of the objective facts regarding the toxic pesticides used in Vietnam.

The government and DVA can spin the data and chant all they want, but it will not change the material facts, one of which is that, without a doubt, in-country veterans were exposed to the deadliest, most organically damaging chemicals in this world. While the truth of the facts is subjective and can change from person to person, the evidence regarding Vietnam veterans is a hard-core fundamental, rooted in scientific and medical reality, and in all likelihood, it will remain constant. The chants and spin being produced by our administrative leaders and DVA is more an advertising campaign, more hoopla than facts. The facts of Vietnam are undeniable, verifiable realities that are rooted in evidence. They shouldn't be matters of opinion. In the end, it all depends on the way you use the wood to build your house.

Now that we have truth and facts straightened out, let's move on to what presumptive probabilities are and why they are needed in the case of veterans who served in Vietnam.

What Are Probabilities, and Why Do Veterans Need Them?

There are two parts to this section, so let's tackle the "What?" portion first. As defined by Congress, a presumptive probability—the type affecting Vietnam veterans—is as follows: an association between the occurrence of a disease in humans and exposure to an herbicide agent shall be considered to be positive for the purposes of this section if the credible evidence for the association is equal to or outweighs the credible evidence against the association.[1]

As noted by Congress and in most legal proceedings, there is wide-ranging agreement about what a presumption is, although there's still considerable controversy about what a presumption does. Perhaps the best and fastest way to explain what a presumptive probability is supposedly designed to do is to look at the opinion given by the National Academy of Medicine:

> Presumptions have played an important role in both the conceptual basis for service connection and the actual administration of the Department of Veterans Affairs (DVA) compensation program. Presumptions are used to *bridge gaps in scientific and medical knowledge,* as well as to resolve complex policy questions and *simplify determinations* of service connection for the VA. A true presumption affects the burden of proof. For instance, a veteran who relies on direct proof to show a service-connected disability must generally, (1) produce evidence on the issue, and (2) persuade VA that the service connection exists. Presumptions are created for a number of reasons. They promote fairness by *simplifying* proceedings and by making *it less burdensome for claimants to gather evidence that is more accessible to the party against whom the claim is asserted.* When the probability of the presumed fact's existence is high if the basic fact exists, presumptions eliminate the expense and time that would be required to establish the presumed fact by direct evidence.[2]

As I interpret the NAM's declaration, a presumption is ultimately a procedural legal device designed to relieve us veterans of the burden of proving actual exposure to harmful pesticides while in Vietnam. But what happens once it is established that we were, as an objective fact, exposed to dangerous toxic herbicides and insecticides in Vietnam that were both genetic and cell-damaging to our bodies?

Stay with me now; this is going to get a little prickly. Once it is proven that we were exposed to numerous toxic tactical pesticides during our tours in Vietnam, the existence of cellular and genetic damage occurring to our internal biological systems on a primal cellular and genetic level must also be presumed, unless the actual exposure evidence is refuted.

Once you get past the organized governmental advertising campaign, the most depressing part of this whole pesticide dispute rigmarole is that if—and it's a big if—our administrative and military leaders had performed their required tasks prior to the war before they knowingly released all their injurious chemicals, we wouldn't be arguing over which illnesses were probably caused by exposure to all the detrimental pesticides. We would know exactly how they affected our health and that of our children.

Similarly, had our government and military leaders, after the war, completed continuing studies of veterans who were exposed to those chemicals and conditions, we wouldn't be arguing over which illness or disorder was probably caused by that service. However, the "precautionary principle" ethics and "reasonable person" standards were never applied to veterans.[3]

As a veteran who was stationed in South Vietnam, I do understand the concepts behind the use of pesticides there, as any reasonable person might. Quite simply, insecticides were deployed in an effort to minimize diseases borne by mosquitoes and other insects, while herbicides were used to deny food and jungle cover to the enemy.

Our legislative and military leaders always had a way of making the war sound so plausible and straightforward. All the same, while the stated theory behind the use of herbicides and insecticides might have been simple and even well intentioned, the purposes for using them were derailed by the lack of research before and after implementation. Likewise, it is hard for me to comprehend how chemical companies and our leaders could have allowed these health-damaging, contaminated pesticides to be used in Vietnam while reasonably knowing that the chemicals contained in their finished creations were systemically damaging, cell-altering, and dangerous to the health of any human exposed to them. That's the part I just don't understand. Nor do I understand the continual denials by our government and their relentless chant of "inadequate or insufficient evidence."

The Mantra: "Inadequate or Insufficient Evidence"

Now for why we need presumptions. Despite the previous simplifying assertions of the NAM concerning presumptive probabilities, there is nothing simple about them. First, we must look at the direct failure of our government and military leaders to act in a responsible, precautionary, and accountable manner. Then add to those failures the vast biological complexity caused by our numerous exposures to some of the most harmful and toxicologically vicious organic chemicals known to modern man, and voilà—the overwhelming need for presumptions.

Our governmental entities (the DVA, NAM, and NAS) have defined the term "inadequate or insufficient evidence" to mean that the available epidemiologic studies are of insufficient quality, consistency, or statistical power to permit a probability conclusion greater than 50 percent regarding the presence or absence of an association between exposure to pesticides and the resulting illnesses. Our administrators make it sound pretty candid and up front. But if we were to take a closer look, just what evidence are they examining?

The fact is the DVA, NAM, and NAS have been ever-so-carefully crafting their presumptive risk assessment probabilities using mostly civilian occupational and workplace exposure studies to principally one toxic chemical—dioxin-TCDD.[4] Our government's risk-assessment probabilities are not being built on the factual data of actual boots-on-the-ground veterans and their experiences. The affirmation and realization that the DVA's risk calculations aren't even based on real-life veterans are quite disturbing, but I have saved that issue for the final chapter of this book.

Consequently, if we were to look at all the objective facts, taking into consideration our extensive chemical exposures and the exceptional conditions under which we served, you would have to conclude that there are absolutely no civilian studies that can replicate the abhorrent conditions we faced in Vietnam. Even more troubling is the fact that our government has not researched or considered the negative health consequences of our multiple exposures and synergistic interactions with some of the most biologically destructive organic chemicals on this planet. The simple fact is that they haven't even included all the hazardous chemical ingredients found in the pesticides used in South Vietnam in their probability assessments. An excellent illustration of this limited research is summed up in our government's very own words:

> Consistent with its prior reviews, NAS concentrated its
> review on epidemiologic studies to fulfill its charge of
> assessing whether specific human health effects are

associated with exposure to at least one of the herbicides utilized or to a chemical component of herbicides, such as TCDD (2,3,7,8-tetrachlorodibenzo-p-dioxin; referred to as TCDD to represent a single—and the most toxic—congener of the tetrachlorodibenzo-p-dioxins, also commonly referred to as dioxin). NAS also considered **controlled laboratory investigations** that provided information on whether the association between the **chemicals of interest** and a given effect is biologically plausible.[5] (emphasis added)

Do you want to know something hilarious? All the time I spent in Vietnam, I never once felt like I was in a controlled laboratory setting. Not even for a second. Go figure.

When I first considered what the combined real-life credible responses by our government and the DVA should have been, I thought, "Exactly what would have happened if our government had studied in-country veterans and all the chemicals, circumstances, and stresses of Vietnam?" After a lot of contemplation on my research and the objective facts, I ultimately had to accept that the only legal, logical, and rational outcome would have been that our government and all their many agencies would have had to admit unequivocally that all our diverse cancers and all our immune, respiratory, cardiovascular, neurological, and endocrine illnesses and disorders were presumptive of our pesticide exposure and service in Vietnam.

Wow, what a concept! Can you imagine the uproar and turmoil a declaration like that would have triggered back in the 1980s or even today? While the above statement is factual and based on evidence, there is no need to worry. Even today, it still appears that the massive financial implications and manipulation by executive administrators will continue to ensure that this significant declaration is not going to happen.

Our government continues to limit the chemicals of interest (COI) being considered and has consistently refused to conduct unobstructed studies of real-life in-country veterans. The fact is, no one even knows how many of the three million or so military personnel who served in Vietnam are still alive today, nor do we know the condition of their health. The government blames this failure on lousy record keeping, but I'm not so sure that's the reason. A bad filing system is one thing; losing track of so many military personnel who were exposed to classified toxic chemicals is different and akin to an outright conspiracy. The government's chant of "insufficient evidence" is not a fact. It is a forceful effort to promote their opinion as fact.

So even though the objective facts are on the side of in-country veterans, we must still find and calculate the probabilities that our exposure to pesticides actually resulted in an increased risk of adverse health consequences. Assessing the health risks involved with toxic chemical exposures, like the rest of the unstudied mess made by the Vietnam War, is as complex as it is controversial.

End Notes - Chapter 12

1. "Veterans' Dioxin and Radiation Exposure Compensation Standards; Congressional Findings and Statement of Purpose, Pub. L. 98-542, as amended by Pub. L. 102-4, §10(a), (b), February 6, 1991, 105 Stat. 19

2. Institute of Medicine, *Improving the Presumptive Disability Decision-Making Process for Veterans* (Washington, DC: National Academies Press, 2008).

3. The "precautionary principle" or precautionary approach to risk management requires that if an action or policy has an actual or suspected risk to have caused or of causing harm to the public, (in this case Vietnam veterans), in the absence of scientific agreement that the action or policy is not harmful, *the burden of proof that it is not detrimental, and we were not harmed by the action/policy falls on those taking, or who took, the action, which in this case was our military and governmental leaders.* Similarly, the so-called *reasonable person* standard in the law of negligence is really a creation of legal fiction. The fact is the *reasonable person* is a theoretical person. They are a legal ideal, focusing on how a typical average person, with ordinary prudence, would act in a given circumstance. To be negligent is to act, or fail to act, in a way that causes injury to another person. But no one's perfect and accidents do happen. What divides an everyday accident from an *act of negligence*, is the *standard of care* required in any given situation. By disregarding the proper customary standard of care for a given situation, an individual, company or institution may be found liable for any resulting injuries.

4. The VA/NAS/NAM's COIs have been only (1) the major toxic components of Agent Orange—2,4-D (2,4-dichlorophenoxyacetic acid), 2,4,5-T (2,4,5-trichlorophenoxyacetic acid), and dioxin/TCDD—and (2) only two components of Agent White: the 2,4-D of Agent Orange and picloram (4-amino-3,5,6-trichloropicolinic acid).

5. Federal Register 77, no. 155 (2012): 47924–28. Chapter 13

Chapter 13
Risk Probability Assessment

We know from interrelated scientific research that veterans who served in Vietnam have an increased risk of adverse health effects being produced by the chemicals used during the war, but just how can we calculate the higher likelihood of those organic toxins adversely affecting their health without longitudinal or observational studies? Currently, there are many different procedures used by researchers to calculate risk probability or impact. After reviewing many from the private sector, I finally came across an assessment process used by the EPA to evaluate health risk for pesticide exposures, a plan they developed from the National Research Council's four-step process.[1]

The first step in this process is identifying the hazard presented and examining each particular ingredient as to whether any of them have the potential to cause harm to people and their environment and, if so, under what circumstances. I think I have covered this point with the chemistry and biology of Vietnam, so we can check it off and move to the next step.

The second step is to evaluate the dose response and examine the numerical relationship between the extent of one's exposure and how much exposure would be needed to produce effects (LOAEL). This second step is a little more complicated because of our biological diversity.

However, in most cases, the amount would be less than the size of a poppy seed.

In the third step, we would have to evaluate veterans' actual exposures and examine what is known about the frequency, timing, and levels of contact(s) with all the pesticides sprayed in Vietnam. Unfortunately, this segment is much more difficult because of official reactions and cover-ups of veterans' toxic pesticide exposures over the last half century. While their responses of outright denial were unjustifiable, there were also problems with military record keeping and the general lack of real-time statistical data from either the military or the chemical companies. The bottom line is I did the best I could with this point by piecing together all the available data I could find.

Lastly, we must characterize the nature of the threat and study how well the information supports decisions about the nature and level of the health risk from exposure to the particular pesticide(s) being used.

This last step is the tricky one because it requires a judgment or opinion on the data that has been produced by the other three stages. Even so, in the United States, as in most developed countries around the world, the use of specific harmful chemicals such as TCDD, HCB, and the many other cell-damaging compounds used in Vietnam is absolutely not permitted unless those chemicals are shown beforehand not to increase the risk of death or illness above a specific threshold.

Chances and Probability Assessment, Las Vegas Style

In the chapter 10 analogy, we had an exposure to one particular substance, ionizing radiation, which was capable of causing multiple health problems, illnesses, and adverse outcomes—one primary substance with various health effects. So the chances and biological probability calculations for the numerous health consequences ionizing radiation presents can be identified and predicted by epidemiological studies with some degree of success, provided the necessary exposure parameters and information are available.

Conversely, boots-on-the-ground veterans were exposed to numerous cell-damaging pesticides containing many different injurious chemical substances, without the benefit of knowing our exposure parameters. We have no valid "before" or "after" studies for reference. Nonetheless, a good comparison for our exposures to multiple toxic chemicals would be if someone were to take a handgun or rifle (exposure to acutely toxic/harmful pesticide) and fire a shot at a person who was unprotected and could not move out of the way. What is the probability of the bullet hitting the targeted person?

If the shooter is a good shot (highly toxic chemical), the likelihood is pretty high. If the bullet does hit the targeted person (amount of pesticide absorbed or ingested), what are the chances of it injuring or killing them? It depends on the ballistics and the caliber of the bullet in addition to exactly where on the person's body the round strikes (similar to the LOAEL and lethal dose measurements of pesticide exposures).

Now a second bullet is fired (another exposure to the same or a different toxic pesticide/chemical). What are the probabilities now, after this second round? What are the chances of injury after a third bullet is fired, followed by a fourth, fifth, and sixth shot, and so on? What would the combined assessment probabilities of damage be after all the available rounds were expended toward the targeted individual? Whatever the probability of the first bullet hitting the person, the additional shots would increase the risks of injury and death exponentially, not linearly.

The same sort of exponential increase occurs when you are exposed to low levels of multiple toxic pesticides and their organic chemical and metabolite components. Multiple pesticide exposures also demonstrate a correlated exponential increase in the assessed biological probability of damage to numerous internal biological systems of our bodies. However, some very dramatic differences exist between bullets and toxic organic chemicals.

The most dramatic and noticeable difference is that bullets will create distinct and immediate visible physical trauma. By comparison, toxic chemical "bullets" produce unseen cellular, hormonal, and genetic damage or alterations that are just as harmful, as real, and—over time— as devastating as any bullet. Nevertheless, with herbicides or insecticides, the destruction occurs a whole lot slower. It can be compared to trying, very patiently and systematically, to kill a person cell by cell, piece by piece, then organ by organ, over a period of years or decades—unhurried, uncompromising, very methodical and relentless, but just as lethal as any bullet. So, even though the effects are considerably slower than those resulting from bullets, bombs, or being hit by a truck, they're ultimately just as destructive and health damaging.

135

The initial cellular and hormonal damage and genetic changes initiated by exposure(s) to toxic chemicals, while unseen and unfelt, do occur—just like they do for ionizing radiation exposures. Thus, veterans will have a much greater risk of serious illnesses and systemic alterations. While veterans do have a substantially increased risk (greater than 50 percent) of adverse health conditions, is there an actual relationship—or, as the DVA would say, a medical nexus—between the pesticides used in Vietnam and the biological health impacts of the toxins they contained?

End Notes - Chapter 13

1. EPA's - Assessing Human Health Risk from Pesticides

Chapter 14
Cause-and-Effect Link

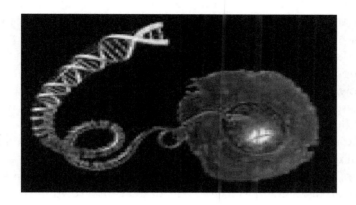

In reality, this chapter is a major premise of this book, to establish a cause-and-effect relation absent the assistance of our legislative leaders undertaking health studies of actual soldiers who were in Vietnam. This one factor, lack of in-country veteran health (morbidity and mortality) studies, makes it very difficult if not impossible, for research groups, veterans, or the DVA for that matter to obtain the information crucial to finding qualitative conclusions about the associations between veterans' pesticide exposures and all the various negative health impacts and outcomes they produced.

As a rule, the primary goal of any epidemiological research or study is identifying an underlying connection between exposure to a harmful substance and health impacts. This association is what scientists call a cause-and-effect relationship or what the DVA would term a "medical nexus." Most scientific causality studies are commonly conducted in a laboratory setting, using animals. While these academic endeavors seem to suggest that establishing a cause-and-effect relationship is easy and straightforward, it is not. In reality, revealing a causal nexus under real-world environmental conditions is one of the toughest and most challenging aspects for any scientific or medical researcher—especially in complex multidimensional situations where there have been many low-level exposures to several toxic pesticides under irregular and hostile environmental conditions.

Practically speaking, isolating and neutralizing the influences of all or even most of the confounding variables of Vietnam is virtually impossible. In fact, it is the multifaceted variables of the war that are the real foundations for all the various cancers, illnesses, and disorders being produced in the first place. The problem isn't that we were exposed to one or even two chemicals. We were exposed to a genuine mélange of highly dangerous, contaminated military pesticides and substances, under conditions that were subprime.

Now if we were to look outside the arena of Vietnam to the physical sciences, such as quantum mechanics or chemistry, we would find establishing causality would be relatively easy, because in any research laboratory, a good experimental design under controlled conditions will neutralize any potentially confounding variables. However, in the case of Vietnam, you can't eliminate the confounding factors and variables; you have to embrace them as a large part of the causal relationship or medical nexus for all the different cancers, illnesses, and disorders being manifested by individual veterans. Of course, while it's very easy for me to spout off and advocate this view—even though there is an excellent factual basis—it does leave us with a uniquely perplexing situation. To be more precise, under what conditions can we use reports, studies, documents, and experimental data to recognize and presume a causal relationship between literally countless variables?

The typical scientific answer would be a resounding "We can't." The reason is—as our government and DVA well know—causality can only be inferred or presumed from well-designed randomized controlled experimentation.[1] But, awkwardly, while this type of testing is satisfying in principle and is sometimes useful in practice, conducting a randomized controlled trial is impossible in this case. Therefore, we find ourselves with the unique, very prickly, unscientific question of whether we can use alternative methods to presume causality from comparable or divergent scientific study and other experimental data. However, to answer that question, I have to fall back on my law enforcement training.

In police investigations, there is a modern trend in motor vehicle accident reconstruction and analysis that accepts the premise that such misfortunes are the result of a number of interrelated causes, none of which can be ascribed as a sole cause. This concept leads to a systems approach of accident analysis, rather than a focus on only one causal relationship.

This process is known as the multi-causality theory. An excellent example is a case I actually supervised. The incident involved a young woman who had a few glasses of wine with her friends at a dinner to celebrate her recent promotion. Walking home from the party, she was hit by an automobile while crossing the street. When officers arrived at the scene, the young woman didn't appear to be seriously injured, and she was refusing medical assistance. The officers reported that she appeared to have only a few minor cuts and bruises. Nevertheless, out of an abundance of caution, she subsequently agreed to be taken by ambulance to a level II regional trauma hospital. A few hours later, the hospital called the investigating officer and told him the accident victim had expired.

During the autopsy, a pathologist attributed the death to a ruptured spleen. The surgeon on duty considered the cause of death to be a slow diagnosis in the emergency room because the spleen could have been repaired had the diagnosis been made promptly. The attending internist attributed the cause of death to exsanguination caused by internal bleeding. The investigating police officer blamed the death on the inattention of the driver, while the driver claimed that the pedestrian was intoxicated and walked right out in front of his vehicle.

Which single cause of death is correct? Each opinion, individually, has some merit, but when they are viewed collectively, an entirely different interactive picture of the circumstances materializes. This example illustrates a simple case of multiple causations in which each of the factors involved could have been the cause of death, and each contributed to the death. Conversely, if the proper action had been taken to prevent the occurrence of any of the events, her death might have been avoided.

Similarly, in the case of those of us who were deployed to Vietnam, each epidemiologist, scientist, or medical specialist tends to base their opinion about our illnesses and disorders being caused by the herbicides and insecticides we were exposed to on their unique allegiance, training, and experiences. Hence, the opinions on which illnesses and disorders were caused by exposures to tactical pesticides would differ noticeably between my primary care doctor and a DVA-certified physician's assistant or between a toxicologist and an immunologist, and so on. These conflicting opinions from different areas of science and medicine are causing some of the misunderstandings surrounding Vietnam veterans.

Our legislators and DVA executives have to get past the invented notion that our numerous ailments are being caused by only one chemical. They are, as a matter of objective facts, being produced by an array of harmful chemicals. It is the combination of all the pesticides and all the circumstances of Vietnam that must be considered collectively rather than separately. Otherwise, the dispute and controversy will continue to rage, and the arguments as to precisely what specific illnesses or disorders were caused by exposure to all the biologically damaging military pesticides will remain unresolved and unanswered—much to the detriment of in-country veterans and their children.

Notwithstanding all the passionately debated controversy swirling around the notion that our cancers and illnesses are not service related (a fact that continues to provoke the ire of many in-country veterans), there is almost no dispute or controversy about the following unbiased facts:

> Fact 1. Deadly dioxin 2,3,7,8-TCDD was contained in the herbicides used in Vietnam—primarily in Agent Orange and the 2,4,-D content of Agent White.

> Fact 2. Dioxin-like compounds—almost as harmful as 2,3,7,8-TCDD—were contained in many military pesticides sprayed and used in South Vietnam, including both herbicides and insecticides.

> Fact 3. Dioxin-TCDD, dioxin-like compounds, and HCB have long half-lives, both in the environment and in humans, with TCDD having a 5.8 to 11.3-year half-life in humans.[2]

> Fact 4. Dioxin-TCDD is one of the most, if not the most, lethal organic chemical in the world, and all the other dioxin-like compounds are not far behind in toxicity and adverse health impacts.

> Fact 5. Extended or multiple exposures to Agent Orange, Agent White, malathion, DDT, and all the harmful chemicals they contained will almost certainly cause serious, widespread damage, alterations, and injuries to the internal biological systems of humans on a primary cellular, genetic, and hormonal level—unseen and unfelt.

On these five points, nearly everyone—other than pesticide companies—agree. What can't be agreed upon is the individual cancers and ailments their interactions produced. Nevertheless, today we have a much better understanding of the extent to which TCDD, DLC, HCB, furans, PAHs, and PCBs truly did harm our physical bodies and internal biological systems, even years after our contacts and exposures in Vietnam had ceased.[3]

In spite of all the complications, this multi-causality approach can work for in-country veterans exposed to the biologically damaging pesticides and environmental conditions of Vietnam. However, the DVA and all the other governmental entities must begin with the premise that our many and various diseases, illnesses, and disorders resulted from a number of interrelated causes and circumstances, with none being attributed to only one reason or one chemical.

Regrettably, we can't go back in time and study the impacts of the pesticides before we were exposed to them. We can't rebuild the data that was lost over the last half century. We can't reconstruct the stressors and environmental conditions. We can't determine all the real genetic and cellular damage caused by the widespread contact with and exposure to the chemicals and adverse circumstances of the Vietnam War. All we can do is presume. Still, presumptive probabilities should not be based on or influenced by partisan politics, costs, or budgetary constraints. Likewise, they should not be constructed on the inferior statistics and study information produced by unsuitable civilian epidemiologic studies.

Civilian versus Veteran Studies

Our legislative executives have seen fit to use civilian workplace epidemiological studies to determine the presumptive probabilities of our cancers and illnesses being the result of service in Vietnam. While it's thought provoking that they chose that path, there is an enormous difference between civilians and boots-on-the-ground veterans, not the least of which is the mere fact that we were living in the middle of a war zone and an intensely hostile environment—and that's putting it mildly.

In the majority of civilian or occupational exposure studies, only one or two chemicals were investigated. By contrast, we lucky Vietnam veterans were exposed to a genuine assembly of physically harmful, heavily contaminated tactical pesticides and associated chemicals.

Thrown into the bargain was the added stress of there being no front lines evident during the war. While the threat posed by the prospect of actual combat was always present, it didn't end there. We were subjected to rocket and mortar attacks, mined roads, and snipers—and if that wasn't bad enough, we were exposed to all the other physical dangers that came just from living in Vietnam.

We met up with hazards on base, in our bunks, and during routine patrols—deadly dangers, such as king cobras and green tree vipers. Then there were the hordes of insects Vietnam had to offer. Just to reiterate a little, the vermin we had to deal with in Vietnam ranged from fire ants, gigantic spiders, and scorpions to rats, fleas, bedbugs, lice, and countless other tropical insects too numerous to mention. In addition to all the pests, snakes, and other ill-disposed critters were the constant threats posed not only by the VC but also by illnesses such as hepatitis, plague, rabies, dysentery, and good old-fashioned jungle rot and by the flood of parasites, such as leeches and liver flukes. Now I ask you, how many civilian study subjects ever had to cope with all those perplexing variables and stresses in their everyday jobs?

We also had the physical stress of being operational seven days a week, twelve hours a day, and sometimes longer. Healthy or sick, we worked. Feel, if you can, the anxiety that we felt as young nineteen-year-old kids, halfway around the world in a strange country, smack-dab in the center of a war. Contemplate the stress we experienced in a tropical/monsoon jungle climate with less-than-optimum sanitary conditions—all without our family or social support systems.

By contrast, the subjects of civilian studies typically lead relatively normal lives in known environments, with their family and social support systems intact. It's entirely unrealistic to believe that our government and the DVA are genuinely acting in good faith when they try to compare the consequences of our chemical/pesticide experiences to those of civilian workplace exposures to only dioxin-TCDD or any of the other very limited chemicals of interest being studied or reviewed by the NAS/NAM. At best, it is inconsistent, and at worst, it is an insult to all the veterans who served in Vietnam.

We lived and were operational in the midst of intense, hostile conditions that merged many diverse, intricate factors. We quite literally showered in a chemical consommé. We drank it. We ate it. We lived with it. We were repeatedly exposed to conditions in Vietnam that are not

142

replicable in any way, shape, or form. There are absolutely no civilian or occupational studies that can even come close to reproducing the chemicals, distasteful conditions, and stress of Vietnam, especially from the early 1960s through the 1970s. Absolutely none!

In spite of all the facts to the contrary, our legislators and the DVA are holding steadfastly to their civilian studies and the marketing campaign they publish in their Agent Orange reports and subsequent updates. In fact, their advertising activities remind me a lot of how grand juries function in the criminal justice system.

Enter the Humble Ham Sandwich

Now stay with me here. I know the heading sounds a little crazy, but I would be willing to bet that most of you have no idea how a grand jury operates. So I will try to convey the link between a ham sandwich and a grand jury.

Traditionally speaking, the constitutional purpose of a grand jury is to provide an impartial method for starting a legitimate criminal proceeding against folks alleged to have committed a felony. With that being said, here is where the ham sandwich comes in. During my law enforcement career, we had an aphorism that a good prosecutor can get a grand jury to indict a ham sandwich. While a lot of individuals use the adage, it's believed to have originated with a chief judge of the New York Court of Appeals, Judge Solomon Wachtler.

It seems that Judge Wachtler believed our grand jury system operated more as a prosecutor's tool than as a protection for ordinary citizens as our founding fathers had intended. Thus, he came up with the ham sandwich correlation. Well, so much for being impartial.

While getting a criminal indictment is easy, proving a case at a criminal trial, in which each side gets its day in court, is an entirely different matter. The fact is, I used this ham sandwich mantra countless times as a police academy instructor. I used it to express the legal concept that any decent prosecutor can get anybody indicted for pretty much anything, principally because the defendant usually has no representation during grand jury proceedings.

Similarly, this simple ham sandwich correlation is an appropriate comparison for our government's Agent Orange newsletters and updates. After reading through many of these DVA/NAS information sheets, it's evident—to me at least—they have forgotten that we were exposed to and came into contact with several of the most biologically menacing chemicals in the world as well as the synergistic impact of others. Thus, the many supposedly "flawed" studies that the DVA/NAS have noted time and again in their reports and updates, while not 100 percent conclusive, are still well over the preponderance-of-credible-evidence (greater than 50 percent) standard and should not be dismissed as having inadequate or insufficient evidence.

The simple reality is the DVA/NAS explanations and opinions published in their newsletters are left unopposed. As a result, you might ask, as I did, "Why are they left unchallenged, and why are their opinions so one-sided?" They're unchallenged because these organizations (DVA and NAS) are believed—at least by our legislative representatives—to be the ones with the medical expertise to make those determinations. Plus, it doesn't hurt that the US government is big and ultimately able to control what its agencies publish. The plain fact is, our government officials can intimidate or even cajole, if and when necessary—just like your everyday "good" criminal prosecutor can.

The unhappy reality is there are no independent medical professionals, autonomous toxicologists, or environmental health specialists fact-checking and scrutinizing the government opinions. There is no one looking at the logic they are applying to the probabilities of our diverse cancers and other illnesses occurring as a result of the pesticides we were unprotected from in Vietnam. No one is checking up on the government's conclusions and opinions. There is no one on the other side of the table representing veterans and opposing the government during those determinations. The next thing you know, our government will be given free rein in trying to indict ham sandwiches for attempted murder because their salt and fat contents are too high

End Notes - Chapter 14

1. A well-designed randomized controlled experiment is one where the participants are randomly allocated into two groups. One group being the experimental group and the other being the control group. As the study progresses, the only expected difference between the two groups is the outcome of the variable being studied.

2. National Institutes of Health's - Report on Dioxin, Agent Orange

3. The National Academies of Science - *Veterans and Agent Orange: Update*

Chapter 15
Our Government's Opinions

As I discussed in chapter 12, a fact is a declaration that can be supported by the objective evidence. In contrast, an opinion is an expression of one's feelings or thoughts that may or may not be based on discernable facts. Indeed, many opinions are based on money, emotions, history, or other tangible values, all of which may not be supported by unbiased evidence.

For example, you may hold the opinion that Agent Orange was more harmful than Agent White. But that does not change the fact that they were both biologically damaging to veterans. Thus, an opinion's value depends on how you reached your conclusions. So how are administrative agencies reaching their opinions?

When the DVA or any other agency tells us that there is insufficient evidence, we must try to understand on what they are basing their opinion. Is it based on measurable data that lead to a convincing conclusion? Or is it based on costs, budgetary restraints, or just an attempt to defend the shameful conduct of past leaders against assertions of chemical warfare?

The following quotation is from the 2010 *Agent Orange Update* published by the NAS. It is just one straightforward example of the judgments of our executive agencies and my contrasting thoughts on their views. Of course, my assessments are different than those published by the NAS. All the same, please keep in mind that it is only one example out of the many I had to choose from:

The NAS noted that two new studies reported statistically significant evidence of an association between herbicide exposure and COPD. A study of Army Chemical Corps veterans reported a statistically significant high mortality rate from COPD. However, the NAS found the significance of that finding to be significantly constrained by the inability to fully control for cigarette smoking, a major risk factor for COPD. The NAS noted that prior studies of American Vietnam veterans did not find evidence of increased *mortality* rates because of noncancerous respiratory conditions. The NAS noted that concerns regarding the misclassification of COPD on death certificates and misdiagnosis of COPD further limited the conclusions that could be drawn from such mortality data. Another new study found a statistically significant increase in the self-reported incidence of emphysema and bronchitis, which are conditions that are consistent with COPD, among Australian Vietnam veterans. The NAS noted that this finding was limited by recall bias and other methodological considerations and expressed general skepticism about the significance of this study's findings because of its low response rate and the study's nearly uniform findings of statistically increased prevalence for nearly 50 health conditions. The NAS further noted that prior studies of the full cohort of male Australian Vietnam veterans showed no indication of increased mortality from COPD or other noncancerous respiratory conditions and that a number of occupational studies failed to detect an increased risk of COPD or other noncancerous respiratory conditions.

Accordingly, the NAS found the evidence to be overall inadequate or insufficient for determining whether an association exists between herbicide exposure and COPD or other noncancerous respiratory conditions. Regarding immune system disorders, the NAS noted that the only potentially relevant new study was the above-referenced Australian veteran study, which found that several conditions in which immune function may play a role—including infectious and parasitic diseases, respiratory disorders, and skin disorders—were significantly more prevalent in Australian Vietnam veterans based on their self-reports than among the general population. For the same reasons discussed above, the NAS found generalizability of the report to be significantly limited by numerous methodological concerns. Accordingly, the NAS found that there was inadequate or insufficient evidence to determine whether an association exists between herbicide exposure and immune system disorders.[1]

In this lengthy selection, the NAS opines—as it has time and time again in its mandated reports—that the evidence is inadequate or insufficient, in its estimation, to warrant any definitive—or proof beyond a reasonable doubt—associations between herbicide exposure and lung and immune system disorders. The commentary and interpretation of the data, though interesting, is not without contradictions or rebuttals. Several key points of contention can be raised regarding the factual accuracy of the comments the NAS makes in just this one 2010 *Agent Orange Update* citation.

The NAS notes, "A number of *occupational* studies failed to detect an increased risk of COPD or other noncancerous respiratory conditions" (emphasis added). As I have already noted, the NAS is trying to compare the consequences of the exposure of actual boots-on-the-ground veterans to inadequate civilian epidemiological studies.

What is more important is that the NAS doesn't note what studies they are referencing or the year(s) in which the studies were completed or by whom. The following is just a short list of studies that do find a correlation between pesticide exposure and COPD and other lung damage:

1. Bertazzi, P. A., Brambilla, G., Consonni, D., & Pesatori, A. C. "The Seveso Studies on Early and Long-Term Effects of Dioxin Exposure: A Review." *Environmental Health Perspectives,* 1998.

2. Chiba, T., Chihara, J., & Furue, M. "Role of the Aryl Hydrocarbon Receptor (AhR) in the Pathology of Asthma and COPD." *Journal of Allergy.*

3. Cypel, Y., & Kang, H. "Mortality Patterns of Army Chemical Corps Veterans Who Were Occupationally Exposed to Herbicides in Vietnam." *Annals of Epidemiology,* vol. 20, 2010.

4. Marshall, N. B., & Kerkvliet, N. I. "Dioxin and Immune Regulation: Emerging Role of Aryl Hydrocarbon Receptor in the Generation of Regulatory T Cells." *Annals of the New York Academy of Sciences*, vol. 1183, 2010, pp. 25–37.

5. OEHHA. *Health Risk Assessment of Malathion Coproducts in Malathion.* May 1997. [Environmental monitoring for malathion, malaoxon, and five malathion coproducts.]

6. Suskind, R. R., & Hertsberg, V. S. "Human Health Effects of 2,4,5-T and Its Toxic Contaminants," *JAMA,* vol. 251, no. 18, 1984, pp. 2372–2380. ["Pulmonary function values among those who were exposed and who currently smoked were lower than those who were not exposed and who currently smoked."]

7. Yi, S. W., Hong, J. S., Ohrr, H., & Yi, J. J. "Agent Orange Exposure and Disease Prevalence in Korean Vietnam Veterans: The Korean Veterans Health Study." *Environmental Research,* vol. 133, 2014, pp. 56–65.

8. "Close attention should be paid to the emerging toxicological literature on malathion coproducts. This is especially important concerning the trialkyl phosphorothioate class of coproducts because of the

nature of the toxicity associated with exposure to these compounds (i.e., changes in lung morphology and fetal deaths in experimental animals)."[2]

While the NAS states that there were no studies supporting the COPD claims of veterans, the studies mentioned above do support such a connection. Moreover, the NAS failed to find those studies, and I am pretty sure they had access to the same studies and data.

Next, the NAS finds cigarette smoking to be a confounding factor in the Army Chemical Corps studies; however, this must be considered a constant. If we examine the influence of cigarette smoking, a well-substantiated fact is that in both the military services and civilian life in the 1960s and 1970s—and most of the 1980s—if you weren't smoking, you were usually exposed to secondhand cigarette smoke, which we have come to learn is as bad or worse than outright smoking.

As anyone old enough remembers, in the 1960s and 1970s, cigarette smoking was allowed in all public places, such as airplanes, buses, trains, business offices, barracks, mess halls, restaurants, clubs, hospitals, and even doctors' offices. Indeed, according to the CDC, in 1965, it was estimated that a whopping 42.4 percent of adults living in the United States were active cigarette smokers. Such a large number of people smoking in so many public places brought about another health concern: secondhand smoke exposure. Consequently, the CDC also published findings on secondhand smoke in the US, which found that between 1988 and 1991, approximately 87.9 percent of nonsmokers in the US had measurable levels of nicotine's metabolite (cotinine) in their bodies due to the effects of hand-me-down smoke exposure.

Very few, if any, no-smoking areas existed in the 1960s and 1970s, and as noted, almost half the population of the country smoked, and 90 percent of those who didn't smoke were still affected by it. These statistical factors would make the influence of cigarette smoking noted by the NAS a constant rather than a confounding variable.[3]

As we continue with the report, one of the studies the NAS comments on compares actual veterans with a population not exposed to Vietnam, and they conclude that the veteran population had a higher incidence of COPD. The study itself contains statistically significant evidence, but the NAS considered cigarette smoking to be a confounding variable, not a constant. It appears from its statements it either overlooked or did not

151

consider the fact that almost everyone, both in the service and civilian environments, was exposed to cigarette smoke, both primary and secondary, in the 1960s and 1970s. The NAS also fails to include or consider the synergistic impact of exposure to herbicides and insecticides or any of the other air pollutants prevalent in Vietnam on the Army Chemical Corps's study subjects.

What is even more troubling is the fact that respiratory cancers—including cancers of the lung, larynx, trachea, and bronchus—are now presumptive Vietnam veteran illnesses, but COPD is not. As noted earlier, the NAS found that a credible study was "significantly constrained by the inability to fully control for cigarette smoke, a major risk factor for COPD." Then again, cigarette smoke is also the number one risk factor for respiratory cancers. This point is a substantial inconsistency in the logic the NAS is applying to COPD in its probability determinations.

Oddly, the NAS did not consider the fact that boots-on-the-ground veterans, in addition to cigarette smoke, were also exposed to almost-constant air pollution and irritating smoke from such things as the burning of areas already sprayed with an herbicide (e.g., Pink Rose), the thick black cloud of smoke wafting up from dung- and trash-burning pits, or the synergistic interactions of inhaling BTEX and diesel-fume smog before publishing their findings. Nor did they consider the fact that the Army Chemical Corps's civilian counterparts were not exposed to Vietnam's environmental conditions.

Instead, the NAS dismissed the results of a study analyzing actual boots-on-the-ground veterans because of the noted confounding factors and their concerns regarding death certificates. Here again, the NAS was looking at and commenting on the increased mortality and not the increase in the morbidity of COPD in veterans. Statistically speaking, most veterans—or civilians for that matter—who have COPD will die *with* their COPD rather than *from* it. This reasonably thought-provoking fact about COPD was noted in a 2008 study published in the *Journal of Respiratory Medicine* and is a statistically significant point that the NAS did not recognize or consider in its probability assessment.[4]

Once again, there were no air quality studies done in Vietnam, so we must look elsewhere for comparative studies and information. The following excerpts are from a now-declassified army memo regarding air quality and environmental conditions at Bagram Airfield, just north of Kabul, Afghanistan. (The war is different, but the pollution is very similar.)

The burn pit at Bagram Airfield produces large quantities of dust and burned waste, which are most likely to impact veterans' health for the rest of their lives, the memo continues. Breathing in the air at the base, veterans risk acquiring any of the following: reduced lung function or exacerbated chronic bronchitis, COPD, asthma, atherosclerosis or other cardiopulmonary diseases. The assessment is built on an eight-year study of weekly air samples taken at the base. The memo says that the air concentration of Particulate Matter 10 and Particulate Matter 2.5 [10 and 2.5 being the diameter of matter in micrometers] were twice and three times the maximum permitted standards, respectively. To soldiers burning anything from broken furniture to human waste in the best of U.S. practice, the hazard comes under a more common name: dust. The air is heavy with it and its "running mate"—the acrid smell of grilled trash.[5]

This recent US Army memo outlines the very same conditions we faced in South Vietnam. However, we never had an eight-year study of weekly air samples done at Cam Ranh, where we had both large army and air force bases from 1966 through the 1970s. While we had many of the same problems noted in the memo above, we also had operations such as Project Pink Rose as a bonus. As many veterans can attest to, the stench of human excrement being burned in a diesel fuel mixture combined with smoldering trash hung over our cantonment area almost daily, which was only compounded by the humid, tropical conditions. Even with all the similarities between Cam Ranh and Kabul, I could not find a single study to determine how those burning operations affected the health of Vietnam veterans. Not one.

Even so, there is still plenty of objective data on the health impact of air pollution in other areas of the globe such as Los Angeles, California, and Beijing, China. As a result of this research, there is little doubt in the scientific or medical community that persistent exposure to burning substances like we had in Vietnam will produce chronic ailments, most of which affect the lungs. Numerous studies have concluded that people exposed to these types of emissions are more likely to develop chronic respiratory conditions such as bronchitis and emphysema.[6]

153

Similarly, an entirely different study presented evidence that the activation of the aryl hydrocarbon receptor by TCDD can modify lung genes programmed for inflammatory responses and mucus production. The end result of these cellular alterations is consistent with a variety of lung diseases such as bronchitis, asthma, small-airways disease, and lung remodeling or fibrosis. They also substantiate the negative role of TCDD activation of AhR in the development of lung damage.[7]

Our government's narrow focus on civilian occupational studies and their restrictions on the chemicals being studied has missed the larger picture of the conditions that were unique to boots-on-the-ground personnel. The government's presumptive assessment probability system just keeps reminding me over and over again of a grand jury ham sandwich structure.

End Notes - Chapter 15

1. Federal Register Volume 77, Number 155

2. Pesticide and Environmental Toxicology Section - Office of Environmental Health Hazard Assessment California EPA- *Health Risk Assessment of Malathion Coproducts in Malathion-*

3. Center for Disease Control and Prevention - *Secondhand Smoke (SHS) Facts*

4. National Academies of Science-*Veterans and Agent Orange: Update 2012 (2014)*– Page 897

5. Military.com – Memo: Afghan 'Burn Pit' could Wreck Hearts, Lungs - Published 22 May 2012

6. "Human Health Effects of 2,4,5-T and Its Toxic Contaminants," *JAMA* 251, no. 18 (1984).

7. Ibid

Chapter 16
DVA's Opinions of My Illnesses

Up until now, you have heard a lot about what I have to say about my ailments and their causes, but what does the DVA have to say about my laundry list of illnesses?

First and foremost, veterans are being required to prove to the DVA that they were actually in the country of South Vietnam. I know it is hard to believe, but according to the paperwork I received from the DVA, they do not know which veterans were stationed in Vietnam and which ones were not. Suspiciously, they claim their records only indicate whether or not the vet served during the era of the Vietnam War. Thus, they assert the veteran must prove they were really in Vietnam when filing a claim for an illness related to pesticide exposures.

As crazy as that requirement appears, they then make a distasteful situation even worse. Once the DVA is satisfied you were really in the country, you must submit evidence that you have one of the few illnesses the DVA presumes is connected with your service there. If your particular illness is not on their presumptive probability list of ailments, the DVA requires you to substantiate that the multitude of harmful pesticides you were exposed to in Vietnam injured you and caused your particular illness or illnesses.

Now, remember here that our government has known for decades that the chemicals used during the war were highly toxic and harmful to humans. We also know the DVA is using civilian occupational studies to determine the ailments plausibly produced in the military personnel who served in Vietnam.

As a result, since reinstituting my disability claims, I have had a number of compensation and pension medical exams at the Lebanon VA Hospital in Pennsylvania. In addition, the DVA requested documentation from me to verify my claims. In response, I have sent the DVA all the studies and data contained in this book, plus even more evidentiary material not listed here to substantiate all my disability claims. In total, I have sent the DVA hundreds of studies and a thousand pages of documents.

So far, the DVA has acknowledged and accepted about half my illnesses as being related to my service; the other half they have rejected. The justifications the DVA gave for rejecting half my illnesses were based on the medical opinions of their certified physician assistant (PA-C). That is correct; you read that properly. The DVA did not even use actual MDs or toxicology experts in making their determinations.

From June of 2012 through September of 2015, I was actively engaged with the DVA, wrangling over the causes of my health problems. So although it was a long process, I will attempt to condense my three-plus years of battling into a brief summary. As noted, the DVA requested a few of their physician assistants to examine me and provide them with their "medical opinions" regarding the probability of my disputed illnesses and disorders being initiated by my service in Vietnam.

M.E.N. is one of the physician assistants who purportedly examined me. Although he did more talking to himself than examining me, he stated the following in his report on my medical illnesses and disorders:

> The veteran was diagnosed with essential hypertension in the 1980s and has been well controlled on pharmacologic agents since then. *Hypertension is not recognized as associated with Agent Orange exposure.* The veteran was not diagnosed with hypertension until the 1980s, not diagnosed while in the military 1965 to 1969.

My response to the DVA was:

M.E.N. should have stated: "This case of hypertension has been well controlled on pharmacologic agents, as well as diet and exercise since then." Nonetheless, numerous studies, medical journals, and medical research professionals would fiercely disagree with M.E.N.'s assessment.

Many medical professionals do recognize a medical nexus between exposure to *Agent Orange, et al.* and hypertension. All the same, the preponderance of evidence from studies and medical journals clearly supports the conclusion that a medical nexus exists between my multiple pesticide exposures in Vietnam and my physician-diagnosed hypertension.

I would also state, for the record, that M.E.N. failed to note in his report—as I had explained to him—that when I left the service in 1969 (in my early twenties), my blood pressure reading was noted as 140/80, and my pulse was 96 BPM. So maybe I should have been diagnosed or at least checked for hypertension by the army, but I was not.

Further on in his report, M.E.N. stated that GERD is not recognized as associated with Agent Orange exposure and therefore is most likely not because of Agent Orange exposure.

My response to the DVA was:

The GERD M.E.N. noted was the result of, and secondary to, my *esophageal achalasia* and subsequent balloon dilatation procedure. Regarding achalasia, I have already addressed that illness with the DVA at great length in my previous "Statements in Support of Case," so I will not belabor the points here.

Next, M.E.N. stated that ulcerative colitis is not recognized as associated with Agent Orange exposure and hence is most likely not an effect of Agent Orange exposure in the military.

My response to the DVA was:

> As I have documented, ulcerative colitis (UC) is a chronic disease of the large intestine and involves an abnormal response or inappropriate activation of the immune system within the large intestine. The organic chemicals myself and many others were exposed to in Vietnam have an established and medically recognized high affinity for lymphoid tissue and other biological structures within the immune system of humans and animals—even though they have not yet been presumed by the DVA or their PA-Cs as service-related illnesses or disorders.

> Here again, the preponderance of evidence from studies, research journals and medical opinions clearly support the conclusion that a medical nexus exists between my pesticide exposure in Vietnam and systemic damage and alterations being produced within my immune system— unfelt hidden cellular changes that would have "more likely than not" initiated my UC.

In summary, M.E.N. stated:

> None of the above-listed *problems is recognized as associated with Agent Orange exposure* and therefore is most likely not associated with military service or Agent Orange exposure.

D.A.M. is another DVA physician assistant who examined me, and he stated the following in his report on my medical conditions:

> Rationale: Agent Orange exposure conceded. Atrial flutter is not recognized as associated with this exposure and therefore is most likely not related to his military time.

158

Rationale: Agent Orange exposure conceded. The two claimed skin conditions are not known, or by current standards, accepted to be related to this exposure.

I could go on and on, but the point I am trying to make is that the physician assistants who examined me are affirming that because my claimed illnesses or disorders were not presumed by the DVA or did not present in service, it was "less likely than not" (less than 50 percent) caused by my multiple pesticide exposure periods and service in Vietnam.

Not one of the physician assistants or anyone else from the DVA noted any of the probative studies, reports, or documents that I submitted in support of my contested illnesses. They did not dispute any of the numerous medical studies or documents I had submitted. Nor did they give any other substantive reason or factual evidence for their opinions other than they were not presumed by the DVA or did not present while I was in the army.

Therefore, it must be concluded that the numerous evidentiary studies, medical opinions, and documents I submitted were ignored by the DVA and their physician assistants. The DVA never addressed the probative research evidence I provided, nor was its corroborating significance challenged or disputed by the DVA's physician assistants in any way. Now, remember that the DVA never once said in any of their correspondence that I was not exposed to pesticides or that the pesticides were not toxic; they never told me that they did not alter my cellular or genetic structure or that they did not affect my health adversely. They only said they were not presumed by the DVA or did not present in service.

Obviously, the DVA is remaining mute on all the probative data and studies I have submitted. So, based on my physician-diagnosed illnesses and all the factual studies and data I have presented to the DVA, it is legally, scientifically, and medically appropriate to conclude that the origin of any disputed disorder of the immune, cardiovascular, endocrine, respiratory, or neurological systems was produced by, and the result of, multiple exposures to all the hazardous chemicals and circumstances of service in Vietnam.

Likewise, based on that very same factual evidence, it is medically and scientifically virtually impossible for the DVA or their physician assistants to realistically and "in good faith" allocate a probability of less than 50 percent to my disputed illnesses manifesting as a result of the cellular, genetic, or hormonal damage that was almost certainly done to my biological systems by the numerous harmful military-grade pesticides and the conditions which were prevalent during my service in Vietnam.

Finally, after exhausting all my options with the DVA regional office in Philadelphia in September of 2015, I submitted a request for a hearing before the Board of Veterans Appeals, and I have been waiting in line since then. My request evidently has been languishing in what the DVA has termed its Appeals Management Center (AMC). Theoretically, according to the DVA, the AMC was designed to coordinate and speed up the Board of Veterans Appeals processing and hearings. However, many veterans have dubbed the AMC the DVA's black hole—cases go in, but they do not come out.

For a little over three years now, my case has just been sitting at the AMC, waiting to be heard. My dispute with the DVA has created a mountain of paperwork, which has caused the death of a small forest. My files are bulging at the seams with studies and correspondence to and from the DVA. I can only imagine what records the DVA are maintaining on just my case.

Nonetheless, one of the many lessons I have learned in life is that when God gives you lemons, do not complain about the fruit; take them and make lemonade. So trying to live by that adage, I decided to take advantage of the opportunity provided me by the AMC and write this book while waiting for my case to be heard by the Board of Veterans Appeals.

When I started this investigative journey into Agent Orange, I never suspected what I would discover. But I quickly learned we veterans were exposed to much more than just one deadly pesticide. The deeper the exploration took me and the more I saw all the lives which had been taken and damaged by the rampant use of pesticides during the war, the more determined I became to try to set the record right. So while my case waits to be heard, the concept for *Silent Spring—Deadly Autumn of the Vietnam War* was born, and a reluctant writer emerged out of sheer exasperation and sorrow.

So Why All the Hullabaloo Anyway?

Although there have been a mountain of back-and-forth correspondence and many bureaucratic procedures to go through, I have tried to explain briefly what the DVA has said about my laundry list of conditions. So without further ado, here is the complete list of my medical problems and how my ailments break down according to the DVA:

Disorders Accepted by the DVA:
1. Chronic obstructive pulmonary disease (COPD)
2. Atrial flutter (heart arrhythmia)
3. Type 2 diabetes mellitus
4. Peripheral neuropathy left and right lower extremity
5. Peripheral neuropathy left and right upper extremity
6. Sensorineural hearing loss, bilateral
7. Tinnitus
8. Severed right ulnar nerve

Illnesses submitted but rejected by the DVA:
1. Fulminate ulcerative pancolitis with toxic megacolon with complete ileostomy
2. Pyoderma gangrenosum
3. Hypertension
4. Gastroesophageal reflux disease (GERD)
5. Nonalcoholic steatosis of the liver
6. Achalasia
7. Hyperlipidemia
8. Allergic rhinitis
9. Seborrheic dermatitis

Complaints not yet submitted to the DVA:

1. Hematuria, unspecified

2. Benign prostatic hypertrophy without flow obstruction

3. Cholecystitis with subsequent cholecystectomy

4. Cataracts—both eyes

5. Detached retina—left eye

So am I really that genetically unlucky, or did the pesticides follow me home? The bad news is there is a massive amount of evidence that leads to the conclusion they did. From the end of the Vietnam War up until around the 1990s, the US government and DVA had vehemently refused to recognize that any of the military personnel who served in Vietnam could have been harmed by the tactical herbicides they allowed to be used. Our government and DVA knowingly classified, suppressed, and ignored any evidence of that harm. They even went so far as to cover up valid research and tried to conduct bogus studies.

It is almost criminal that political expediency and budgetary constraints have been allowed to eclipse the substances to which in-country veterans were exposed. While it is disgraceful that veterans' health problems have been disregarded over the decades, even worse is the fact our government has ignored and covered up the objective facts about what really happened to the health of in-country veterans and even many of their children.

Chapter 17
Pesticide Cover-Ups Begin

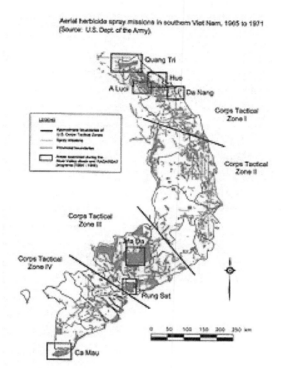

Aerial herbicide spray missions in southern Viet Nam, 1965 to 1971
(Source: U.S. Dept. of the Army).

What was the single greatest conspiracy of the twentieth century? Which one was so ruthless in its influence and so outrageous in its construction that it deserves top billing?

With so many intriguing events taking place in the 1900s, picking just one is hard. Of course, many conspiracy concepts will amount to nothing. But, some conspiracies will be more than just a theory. They will not only grab your attention, they will make your skin crawl and the hairs on the back of your neck stand up. They will provoke you, making you ask, "What really happened?" Although I am sure everyone will have their own favorite scheme, the most shameful conspiracy ever cooked up by human minds in the twentieth century was not the grassy knoll: it was the Vietnam War.

163

You do not have to look too far to find the Pentagon Papers. Their official title was "Report of the Office of the Secretary of Defense Vietnam Task Force." The top-secret report was a forty-seven-volume, seven-thousand-page study ordered by Robert McNamara to get answers to the murky questions swirling around the expanding war in Vietnam. The research he received concluded that Vietnam was an unwinnable war.

As we continue investigating, we learn of the petition from the Federation of American Scientists sent to President Johnson in February of 1967. The appeal contained over 5,000 signatures of scientists, including 17 Nobel laureates, and 129 members of the National Academy of Sciences. The main two points of the request were first, to persuade Johnson to end the pesticide spraying in Vietnam and second, to raise concerns about the ecological impacts of the herbicides being used there.

Moving right along, we discover an internal memorandum by the Dow Chemical Corporation dated February 22, 1965, which provides us with a summary of a meeting of thirteen top executives of the company. The memo's stated purpose was to discuss the highly credible systemic hazards posed by dioxins in their 2,4,5-T product, which was also a significant component in the formulation of Agent Orange. As a result of that meeting, Dow officials decided they needed to have a conference with other makers of the chemical 2,4,5-T and to formulate a united stance—or what I would call spinning the facts—on its dioxin content.[1]

In March 1965, Dow official V. K. Rowe arranged a get-together with executives from Monsanto, Hooker Chemical (which operated the ill-fated Love Canal dump), Diamond Alkali (the forerunner of Diamond-Shamrock), and the Hercules Powder Company, (which later became Hercules, Inc.). According to recently discovered documents, the purpose of this meeting was "to discuss the toxicological problems caused by the presence of certain highly toxic impurities" in samples of 2,4,5-T.[2]

The primary highly toxic impurity was—you guessed it—dioxin-2,3,7,8-TCDD. Excerpts from the now-declassified letter written by Rowe to bio-products manager Ross Milholland on June 24, 1965, detailed the following (to paraphrase): Dow knew, as a fact, that the dioxin-2,3,7,8-TCDD contained in their products could cause serious systemic harm and damage to anyone unlucky enough to be exposed to them, or as Rowe stated in his letter, "2,4,5,-trichlorophenol (2,4,5-T) and TCDD are exceptionally toxic and have tremendous potential for producing chloracne and systemic injury."[3]

Rowe went on to state, "The whole 2,4,5-T industry would be hard hit, and I would expect restrictive legislation, either barring the material or putting very rigid controls upon it." Wow! In addition to being damming, the statement also sounds like Rowe was putting profits before safety. Moreover, it makes you wonder what other sneaky, underhanded dealings went on behind those classified closed-door meetings that haven't been found. In any event, Rowe closed his letter with the following:

> There is no reason why we cannot get this problem under strict control and thereby hopefully avoid restrictive legislation. I trust you will be very judicious in your use of this information. It could be quite embarrassing if it were misinterpreted or misused. Under no circumstances may this letter be reproduced, shown, or sent to anyone outside of Dow.[4]

The Rowe letters were not the only attempts made to whitewash the negative health impacts of the pesticides used in Vietnam. There was also a $43-million study being conducted by the CDC on the health risks posed to military personnel by Agent Orange. However, shortly before the CDC was to release their study results, the Institute of Medicine (now the NAM), submitted a harsh rebuke of the CDC's study test methods. The Institute of Medicine claimed (to paraphrase): none of the CDC's conclusions were supported by scientific data. As a result, the CDC abruptly halted their report in 1987. They subsequently claimed a lack of military records made it impossible to determine which soldiers were exposed to the herbicide and ultimately refused to turn their final report over to the White House.[5]

Then, as we search even further, we find a draft of the Ranch Hand Study, which showed twice as many congenital disabilities among the children of Ranch Hand participants. The draft also confirmed that the Ranch Hand study subjects were not as healthy as the control subjects by a ratio of 5:1 (sick to healthy). However, these facts were somehow strangely deleted from the final Ranch Hand report, which stated that there were no adverse effects observed following exposure to Agent Orange.

Now you might be asking yourself, as I did, "How could that possibly be correct?" Once again, according to declassified records, Air Force study personnel deleted the findings from the final report at the "suggestion" of a Ranch Hand Advisory Committee set up by the White House Agent Orange Working Group (AOWG). It seems that the air force scientists involved in the study were pressured by the air force command and the White House to change the results and delete critical information from the final report.[6]

Another study in the late 1980s was obstructed and manipulated by the White House, primarily through the AOWG and the Office of Management and Budget because the Reagan administration had adopted a legal strategy of refusing liability in military cases of exposure to toxic chemicals or radiation. The most notable portion of the report was the conclusion that the federal government had blocked or downplayed the findings of ill health among Vietnam veterans that could be linked to pesticide exposure.[7]

Regrettably, that is just a small portion of a very large iceberg. God only knows how many other studies were hidden or classified, covered up, lost, or skewed in an effort to control and minimize the negative impacts of the pesticides to which veterans were exposed. We will probably never know the true extent of the plots and misconduct, thanks in large part to contrived national security issues and the burying of many relevant documents in strange places, such as the Department of Agriculture.

Initially, our government classified most of these critical documents for decades. Today, however, many have been declassified and are obtainable; nonetheless, you must find them first. Once you find them, they help confirm the extent of the manipulation, interference, and cover-ups by chemical companies and our legislative leaders. I must also emphasize that none of the whitewashing actions I have noted could have been carried out successfully unless they came from and were supported by the top leaders of our government and military—senior executives and administrators powerful enough to manipulate and cover up all the negative studies and adverse information on the various pesticides to which veterans were exposed.

So why would our governmental executives—including each presidential administration from Richard Nixon in the 1970s to Ronald Reagan in the 1980s to George H. W. Bush and Bill Clinton in the 1990s to George W. Bush, Barack Obama, and Donald Trump in the early part of the twenty-first century—want to manipulate and cover up all this relevant information in the first place? What could their motivation possibly have been?

Unfortunately, the whys are entirely mercenary—the cost of health care and disability benefits for tens of thousands of additional military personnel would be astronomical. The second is to protect the pesticide manufacturers, along with the US government, from international liability claims and chemical warfare issues.

Initially, our government ignored and suppressed all evidence of harm to veterans from any of the military herbicides and insecticides they used during the war in Vietnam. To make matters worse, research was conducted in an effort to try to prove that these harmful chemical substances were supposedly safe—research that we know today was improperly performed, intentionally skewed, and knowingly bogus, thanks in large part to a report submitted by Admiral Elmo R. Zumwalt Jr.

Report by Admiral Zumwalt, May 5, 1990

Our military leaders made multiple command decisions during the Vietnam War to spray millions upon millions of gallons of pesticides that contained several of the most toxic and biologically damaging chemical substances known to modern man. They knew or at least should have known that just the dioxin-TCDD component of Agent Orange and all the other DLCs were extremely dangerous and physically harmful.

It has been estimated from the accessible military records that more than twenty-two million gallons of concentrated herbicides were sprayed in Vietnam. Of that total, 132,596 gallons of Agent Orange and 77,215 gallons of Agent White concentrates were used in at least 272 missions over Khanh Hoa Province, where Cam Ranh is located. Additionally, we know from now declassified records that, at the very least, 1,373 gallons of Agent White and between 4,530 and 22,227 gallons of Agent Orange concentrates were disbursed in thirty-one missions over Cam Ranh itself (and even more, categorized within the data from Khanh Hoa Province).[8] Keep in mind these statistics are only for the two primary herbicides used in Vietnam. They do not include any of the other herbicides or insecticides used there.

Then in the years following the war, our government made the health issues of returning veterans even worse. Despite knowing the toxicity and dangerous nature of the chemicals used in Vietnam, they denied benefits to veterans like Larry who contracted illnesses and other varied medical problems, just because they couldn't prove a medical nexus between their condition(s) and all the classified chemicals they had been unprotected from during the war. Our administrative officials successfully continued that disgraceful policy of demanding that we supply the evidence (nexus) for all our many diverse cancers and illnesses from the end of the war until the 1990s.

On May 5, 1990, everything began to change when Admiral Zumwalt presented his recently declassified report to Edward Derwinski. The official version was titled "Report to Secretary of the Department of Veterans Affairs on the Association between Adverse Health Effects and Exposure to Agent Orange."

Admiral Elmo R. Zumwalt Jr. commanded the naval forces in South Vietnam and was the commander who ordered the spraying of herbicides in Vietnam. What better words could we listen to about the damage done by the herbicides he himself had ordered deployed during the Vietnam War? The following is an actual excerpt from the admiral's report:

> After reviewing the scientific literature related to the health effects of Vietnam Veterans exposed to Agent Orange as well as other studies concerning the health hazards of civilian exposure to dioxin contaminants, I conclude that there is adequate evidence for the Secretary to reasonably conclude that it is at least as likely as not that there is a relationship between exposure to Agent Orange and the following health problems: Non-Hodgkin's lymphoma, chloracne and other skin disorders, lip cancer, bone cancer, soft tissue sarcoma, birth defects, skin cancer, porphyria cutanea tarda and other liver disorders, Hodgkin's disease, hematopoietic diseases, multiple myeloma, neurological defects, autoimmune diseases and disorders, leukemia, lung cancer, kidney cancer, malignant melanoma, pancreatic cancer, stomach cancer, colon cancer, nasal/pharyngeal esophageal cancers, prostate cancer, testicular cancer, liver cancer, brain cancer, psychosocial effects, and gastrointestinal diseases . . .

Such a resolution of the embarrassingly prolonged Agent Orange controversy would be on the order of decisions to compensate U.S. soldiers who contracted cancer after exposure to radiation from atomic tests and U.S. soldiers involved, without their knowledge, in LSD experiments. With the scientific basis now available for it to be stated with confidence that it is at least as likely as not that various health effects are related to wartime exposure to Agent Orange, there is the opportunity finally to right a significant national wrong committed against our Vietnam Veterans.[9]

When deployed in Vietnam, pesticides such as Agent Orange, Agent White, DDT, and malathion were—in theory, at least—intended to save the lives of soldiers who were in active combat situations. Nevertheless, in retrospect, the military's use of heavily contaminated tactical pesticides may have saved some lives on the battlefield, but they ultimately claimed, injured, and destroyed many more lives than their use saved.

I really do understand why the US government, the military, and the various chemical companies do not want to admit liability or study actual boots-on-the-ground veterans. Besides the cost of caring for so many more soldiers, marines, and sailors, there is also the humiliation and responsibility ascribed to the deployment of these exceptionally harmful tactical pesticides and chemicals without doing proper studies or having adequate safeguards in place beforehand. At the least, the Vietnam War was a combination of shameful events that never should have been allowed to occur. At the very worst, Vietnam was a monumental military blunder concocted by inept governmental leaders and administrators.

Regrettably, we can realistically equate the deployment of tactical pesticides in Vietnam to carpet-bombing areas in which your own troops were positioned. This, of course, would put US leaders in a very awkward situation. All the same, unlike exploding bombs, pesticides are not dramatic, and the internal harm they cause is not immediate or visible. The damage and injuries produced by pesticide air strikes would initially be hidden and unfelt. The concealed cellular and genetic injuries would take years to become evident in the military personnel exposed. What more can be said? It was ridiculously easy for officials to cover up the damage caused by the delayed consequences of their pesticide bombings.

The cellular and genetic damage and alterations sustained from pesticide bombardment cannot be quantified or accurately established because of the failure of the US government and military to keep accurate records and to do the necessary research before, during, and especially after our tours of duty. And as if blanketing us with toxic pesticides wasn't bad enough, our legislative and military leaders, along with other DC bureaucrats, almost unanimously agreed that they would not accept the blame, responsibility, or any of the costs for their inappropriate decisions during the war. A point very well made by Admiral Zumwalt during an extensive taped interview he gave on July 26, 1999, where he stated the following:

> We found and reported in that the Bureau of the Budget had ordered all the agencies of government in essence not to find a correlation between Agent Orange and health effects, stating that it would be most unfortunate for two reasons: the cost of supporting the veterans and the court liability to which corporations would be exposed.[10]

How many government agencies do you think were ordered not to find a correlation between Agent Orange and adverse health effects?

We just do not know and probably never will. What we do know is that the National Academy of Medicine admitted in its 2008 *Agent Orange Update* that having the organization evaluate the impact of a combination of toxins on Vietnam veterans was just not feasible, but they never revealed any reasons for this impractical determination. As a consequence, what Admiral Zumwalt said back in 1990 may be equally valid today in 2018: "The cost of supporting veterans and the court liability to which the government and corporations would be exposed" still appears to be a very significant consideration for our leaders. If, however, this was not the case, and the US government really wanted to find out what happened to veterans' health in Vietnam, then I am quite sure legitimate, unobstructed studies of in-country veterans would have already been accomplished—studies that would have resulted in many more cancers and illnesses being added to the DVA's presumptive list of ailments.

Covering-up the Toxicity of Dioxin et al.

While legislative and administrative leaders would downplay the toxicity of dioxin et al., we know factually—just by the extensive evidence cited in this book and from other venues—that dioxin-TCDD, DLC, HCB, and all the other chemicals veterans were exposed to in Vietnam will cause cellular, hormonal, and genetic damage, alterations, and injuries to the primary organs and systems of the human body. However, I can agree on one point made by the DVA: just because something can be, or is known to be, toxic and injurious does not mean that we automatically need to worry about it. Assessing the actual risk of exposure to any individual chemical substance or compound depends on the following three points:

1. The real hazard the substance presents

2. The amount necessary for the material to be toxic—the lowest observed adverse effect level (LOAEL), median lethal dose (LD_{50}), and absolute lethal dose (LD_{100})

3. The exposure time and route of contamination

A simple example at the very low end of the toxicity spectrum is pure water. Yes, that is correct; water can be deadly, even though we depend on it for our health and normal physical functioning. Still, if a person was to drink large amounts of pure water in a short period of time, the water consumed would be injurious because of the intricate relationship water has with the minerals in our bodies. By drinking too much water too quickly, we dilute the balance of electrolytes—primarily sodium and potassium—to the point where our biological systems can no longer function properly. So if you drink too much too quickly, it can kill you.

Obviously, on the one hand, drinking such large quantities of water is not a concern in most cases, so we can presume that the probability of pure water injuring us is very, very low. On the other hand, in terms of toxicity assessment, dioxin-TCDD, HCB, and DLC are at the complete opposite end of the spectrum. With pure drinking water, an amount in the region of six liters would have to be consumed in a very short period to harm a 165-pound person. Conversely, the amount of dioxin-TCDD required to harm that same 165-pound subject would be smaller than a grain of table sugar. Thus, the total avoidance (100 percent) of dioxin is highly recommended by almost every environmental toxicologist.

It is unfortunate that we still have so much to learn about how dioxin-TCDD and other pesticides affected the health of veterans because of all the governmental interference and whitewashes. Nevertheless, one of the critical facts we do know is that there are important differences in how humans accumulate dioxin compared with experimental animals. Several studies have found that in animals (such as rodents), dioxin impacts have a short half-life of ten to thirty days. In stark contrast, the half-life of dioxin is 5.8 to 11.3 years in humans. This longer half-life also accounts for the "internal exposure" phenomenon observed in humans that does not occur in lab animals. Thus, dioxin(s) will accumulate in human tissue at a higher rate and affect our cellular and genetic structures for more extended periods than it does in experimental animals.[11]

The Cover-Up Continues Today

Unlike small animals, humans do not appear to be susceptible to the rapidly fatal effects of dioxin. However, even here the debate is extensive because in a real, live human being an unknown preexisting genetic or medical condition can always be blamed for a particular illness rather than exposure to toxic chemicals. This is especially true in cases where one relatively rapid fatality after another occurs, as would be the case in the aftermath of an accidental event such as the Seveso Incident.

In the EPA's 1992 dioxin reassessment, which was reinforced in the 1994 update, the agency restated the following:

> Cancer may not be the most sensitive toxic response resulting from dioxin exposure. Immuno-toxicity and reproductive effects appear to occur at body burdens that are approximately 100 times lower than those associated with cancer.

As a consequence, even after more than a half century of governmental meddling and scanty scientific study, the topic of just how harmful dioxin and Agent Orange et al. were in Vietnam is still hotly debated. Agent Orange's toxicity is an extremely complicated topic in and of itself, even without considering all the other herbicides, insecticides,

172

and inert chemicals used in Vietnam.

The fact that there are two sides to this heated pesticide argument is no secret. On one side of the dispute are the chemical companies (e.g., Dow Chemical and Monsanto) that produced Agent Orange and other pesticides for military use during the Vietnam War. Even today, they continue to be embroiled in controversy over the herbicides and insecticides they produce and the effects they have on our military and civilian populations.

As a matter of fact, in 2014, a group of thirty-five doctors, scientists, and researchers objected to the use of the chemical we know as 2,4-D by Dow Chemical. According to the details of their letter, this small group sent a request to the EPA recommending that the agency deny the approval request by Dow to revived the use of 2,4-D in a new double herbicide mixture. This new weed killer was eventually marketed under the name Enlist Duo. According to the data, Enlist Duo is a combination of 2,4-D (of Agent Orange infamy) and glyphosate (an organophosphorus herbicide).[12]

Now, if we are speaking about organic chemistry, there are structural differences between organophosphorus and organophosphate pesticides. However, if we are looking at the health impacts of organophosphorus (e.g., glyphosate) and organophosphate ones (e.g., malathion), then they are virtually the same. Both organic substances are acetylcholine inhibitors and have similar effects on the human body.[13]

Unquestionably, persuading regulatory agencies, veterans, exposed residents, and ex-employees that their new (or old) herbicides like 2,4-D and organophosphate insecticides are not harmful by using data from unpublished, improperly designed, or skewed studies is in the best interest of chemical companies. The nontoxic spin is of vital economic importance to these companies today, just as it was back in the 1960s and 1970s—maybe even more so currently because of all the mounting evidence regarding the negative health impacts of their very profitable pesticide products.

On the flip side of the issue are the veterans, employees, and residents who have been exposed to varying levels of dioxin, other toxic substances, and the inert secret ingredients of pesticides. They must deal with the consequences of exposure(s) and the substantial number of illnesses and adverse health effects, including higher death rates, cancers, and numerous other systemic illnesses. All of this is being vehemently denied by pesticide companies because, according to their spin on the facts, there is insufficient evidence to prove otherwise.

The upshot is that all the highly toxic organic chemicals in-country veterans were exposed to will, with very little room for doubt, attack living organisms on an unseen and undetected hormonal, genetic, and cellular level, even when the exposure is in microscopic quantities— amounts of less than the size of a poppy seed. No matter how you try to spin it, these are the objective facts on the unhappy realities of pesticide use. Even so, there are still a lot of illusionists and magic tricks around, trying very hard to alter and twist the real facts.

End Notes - Chapter 17

1. *Files Show Dioxin Makers Knew of Hazards* - by Ralph Blumenthal – NY Times – Dated 1983

2. Dow Chemical Company v. Mahlum - Supreme Court of Nevada No.⬚28600 - Decided: December 31, 1998

3. *Files Show Dioxin Makers Knew of Hazards* - by Ralph Blumenthal – NY Times – Dated 1983

4. *History of Agent Orange Use in Vietnam: An Historical Overview* - United States – Vietnam Scientific Conference on Human Health and Environmental Effects of Agent Orange/Dioxins March 3-6, 2002 Hanoi, Vietnam

5. The History of Agent Orange use in Vietnam a historical overview by Paul Sutton

6. Ibid

7. 101st U.S. Congress 2nd session - House of Representatives - Report 101-672 – Executive Summary: The Agent Orange Cover-up: A Case of Flawed Science and Political Manipulation - 1990

8. Alvin L. Young Collection on Agent Orange Container List - Document # 00108 - *H.E.R.B. Tapes: Defoliation Missions in South Vietnam, 1965-1971. Data by Province.*

9. Department of Veterans Affairs: Report to the Secretary of The Department of Veterans Affairs of the Association Between Adverse Health Effects and Exposure to Agent Orange.

10. Excerpts from a taped interview by Moon Callison with Admiral Zumwalt on July 26th 1999

11. National Instituted of Health - *Report on Dioxin, Agent Orange,* 2,3,7,8-Tetrachlorodibenzo-p-dioxin CAS No. 1746-01-6

12. Environmental Working Group - Doctors, Scientists Urge Obama Administration to Reject Potent Herbicide Mix – by Sara Sciammacco

13. Acute health effects of organophosphorus pesticides on Tanzanian small-scale coffee growers – Journal of Exposure Science & Environmental Epidemiology 4 Sept. 2001

Chapter 18
The Old Smoke and Mirror Illusion

My Wife or Mother-In-Law

The motto at the beginning of several of the studies produced by the National Academy of Medicine is "Knowing is not enough; we must apply. Willing is not enough; we must do. —Goethe." And I could not agree more. During my years of research, I found that the work being conducted by the NAM/IOM on the pesticide exposure(s) of veterans is among the best available. Nevertheless, the NAM/IOM is only as good as the studies and information it has to work with. Time after time, in NAM/IOM presentations, you find the following notation in the section where it lists the study or studies it is reviewing: "Studies of Vietnam Veterans: NONE."

In addition, the NAM/IOM is only studying a select few chemicals of interest; they are not being allowed to consider all the very toxic chemicals and substances found in the real-life, heavily contaminated, military-grade pesticides deployed in Vietnam. To illustrate this point, let's take a look at the chemical called picloram, which the NAM/IOM has studied with respect to the effects of exposure on animals. Now keep in mind that picloram was just one of the harmful synthetic ingredients in the real-life Agent White used in Vietnam. Here is where the "smoke and mirrors" enter into the picture. The picloram being used by the NAM/IOM is the purified chemical; it is not the picloram mixed with 2,4-D and contaminated with HCB or any of the other harmful substances found in the real-world version of Agent White. It is the right chemical, but it is being studied in its purified form, separate from all the other interactive elements found in the pesticides sprayed during the war.

Another excellent example of misdirection can be found in the many substances *not* being studied by the NAM/IOM. By not completely investigating the chemicals of Vietnam they are effectively being kept off anyone's radar and out of discussions. These out of sight materials are compounds such as BTEX, hexachlorobenzene, DDT, and malathion.

Without question, the work the NAM/IOM has undertaken is essential. The information its research is producing is highly enlightening. However, as with other Agent Orange governmental research groups, the question of the impacts of all these other unstudied chemicals is still not being addressed, evaluated, or even considered. Whether the NAM/IOM believes these other substances are irrelevant to its mandate, this lack of attention is just an oversight, or some politics or government interference has crept in, I honestly do not know the reason for not including all the chemicals we were exposed to during the war. What I do know and have acknowledged is that if we consider, investigate, and study all the chemicals and conditions of Vietnam together and jointly (multi-causality theory) instead of purifying and breaking them apart, then a new unabridged, associative, and interactive probability assessment would become readily apparent.

When all is said and done, the ultimate point is that the US government has allowed only a very few chemicals of interest to be studied and then has only allowed them to be considered in their pure, unadulterated, sanitized versions—forms that are not even remotely close to the injurious chemical gumbo veterans were exposed to in Vietnam.

One-Size-Fits-All Government Strategy

Our pesticide exposure and experiences in South Vietnam were unique. They were not a one-size-fits-all problem, and they do not have a one-size-fits-all solution. Take for instance the following DVA excerpt from their *Vietnam Veterans and Agent Orange Independent Study Course* (updated June 2008), where they state:

> At first, [the] DVA had a hard time responding to the questions and health concerns of returning veterans. For one thing, military medical, personnel, and exposure records were not maintained with future epidemiologic research in mind. The inadequacy of these records for research purposes continues to this day to thwart scientific investigation of possible long-term medical consequences among Vietnam veterans from Agent Orange exposure. In addition, in the 1970s and early 1980s, there was far less scientific information available about the long-term health effects of herbicides and dioxins in any exposed population, further complicating [the] DVA's ability to evaluate health problems among Vietnam veterans. Initially, this combination of minimal exposure data and limited scientific understanding of Agent Orange and dioxin health effects left [the] DVA poorly prepared to respond to mounting concerns among veterans and others. These limitations also affected [the] DVA's ability to establish defendable policies on Agent Orange disability compensation. Moreover, the issue of Agent Orange and health and [the] DVA's inability to effectively respond became a lightning rod for those concerned with the fair and equitable treatment of Vietnam veterans.

Although there are many disconcerting declarations in this DVA educational program, two concerns jump out from this short excerpt. The first is that the DVA is only focusing on Agent Orange exposure. The second is that the DVA appears to be far more concerned about its strategy of "establishing defendable policies" than about finding out exactly what exposure to the onslaught of all the nightmarish chemicals did to veterans, their health, and the health of their children.

Judging by the many volumes of defendable policies and procedures established by the DVA, their strategies would appear to have succeeded exceptionally well, although they are very perplexing. While the DVA has succeeded in developing policies and procedures that may be defensible, their rules are not medically or biologically irrefutable, always accurate, or invulnerable. So although the government and DVA have had great success in establishing, monopolizing, and defending their presumptive probability list of illnesses, as well as limiting their international liabilities, they have failed in implementing studies that examine actual veterans.

Consequently, the DVA has been unsuccessful, by design, in recognizing and categorizing the vast majority of cancers, illnesses, and other disorders more likely than not produced by chemical exposures in Vietnam. Regrettably, whether the reason for their failure is political expediency, fiscal conservatism, or administrative acceptability, or even if it is because of international chemical warfare liability concerns, it really doesn't matter. What matters is that the US government and DVA have failed to keep their commitments to in-country Vietnam veterans, no matter how they try to spin the facts of our service. Eventually, the number of illnesses caused by our exposure(s) will turn out to be enormous, despite what the DVA may or may not do or say—or even presume.

More trickery?

Despite over forty years of struggling to evaluate whether pesticide spraying affected veterans who served in Vietnam and the health of their children, the controversy still rages today. As a result of the many still-unresolved scientific issues regarding the health and genotoxic effects of the chemicals sprayed in Vietnam, the DVA has begun to create a new system—or if you will, a new scheme—which it describes as the "systematic review of the epidemiologic and toxicological evidence by Institute of Medicine committees," as a way to synthesize all the evidence and guide them in their decisions on benefits for Vietnam veterans.[1]

But wait! Why are there so many unresolved scientific issues almost a half century after the war? Yep. You probably guessed it. It is because the government has impeded the release of information and has steadfastly refused to study veterans who actually served in South Vietnam.

Complexity and intricacy have been acknowledged throughout this book's exploratory endeavor, and the objective factual evidence offered has a broad and multifaceted base. However, the success of the DVA's synthesis of evidence is not merely a question of the merit of its underlying ideas; it rather depends on the degree of confidence and trust in the individuals, organizations, and infrastructures through which and in which the synthesis of evidence is delivered. If there is little or no confidence or trust in the US government, then the whole idea of integration is for naught.

Trust and accountability are critical requirements in any fusion of evidence. Synthesis should be used to improve the thinking that goes into all the documented pesticides and associated chemicals used in Vietnam and the stress known to have been experienced by the personnel stationed there, especially those in active combat roles. As a result, if used correctly, this process of synthesis will provide an ethical and honorable turn-away from issues that are misleading, moving from the failed one-size-fits-all (only dioxin) way of responding to the many health problems and disorders afflicting tens of thousands of veterans.

Nonetheless, as this investigative work has established, the great strength of the DVA's idea of synthesis is in constructing a representation of how various mixtures of highly dangerous pesticides, chemicals, and circumstances worked together and how they negatively affected the health of veterans and their children.

End Notes - Chapter 18

1. National Organization on Disability - U.S. Vietnam Veterans and Agent Orange: Understanding the Impact 40 Years Later

Chapter 19
What Has Been Established?

The US government has conceded that every veteran who set foot on the soil of South Vietnam was exposed to toxic pesticides. While they have acknowledged contacts, they have not accepted the health costs or responsibilities for those exposures. So what are the facts? What can be proven by the significant medical and scientific data available? What are the factual certainties of the pesticides veterans were exposed to in Vietnam?

Once again, while there is a lot we don't know about how the health of military personnel was affected by the pesticides unleashed in Vietnam, this is what we do know:

1. As confirmed by military records and scientific data, the pesticides used in Vietnam contained at the very least the following known chemicals and compounds:

**2,4-Dichlorophenoxyacetic acid (2,4-D)

**2, 4, 5-Trichlorophenoxyacetic acid (2,4,5-T)

**Picloram (4-amino-3,5,6-trichloropicolinic acid)

**Hexachlorobenzene (HCB)

**Triisopropanolamine

**2,3,7,8-Tetrachlorodibenzo-para-dioxin (TCDD)

**Dioxin-like compounds (DLC)

**Benzene, toluene, ethylbenzene, and xylenes

**Polycyclic aromatic hydrocarbons (PAHs)

**Polychlorinated biphenyls (PCB)

**Volatile organic compound (VOC)

**Dichlorodiphenyltrichloroethane (DDT)

**Malathion - S-[1,2-bis-(Ethoxy-carbonyl)ethyl]-O,O-dimethyl-dithiophosphate

**O,S,S-trimethyl phosphorodithioate (OSS-TMP)

**Malaoxon ($C_{10}H_{19}O_7PS$)

**N,N-Diethyl-meta-toluamide (DEET)

2. By using just basic college chemistry and biology, we can determine that the herbicides and insecticides used in Vietnam had a much more powerful adverse physical impact on personnel when their exposures were combined.

3. Massive amounts of private research and countless studies have verified the great harm, and negative health effects that result from exposures to just the most prevalent chemicals contained in the herbicides and insecticides used in Vietnam. It is a fact that the chemicals used in Vietnam caused systemic cellular injury, hormonal alterations, and genetic damage, all while being unfelt and too small to be seen or diagnosed.

4. The government and DVA are basing their presumptive assessment probabilities of our cancers and illnesses on inadequate and inappropriate civilian occupational workplace research and studies.

5. According to declassified records, of the conservatively estimated 20 million gallons of highly concentrated tactical herbicides sprayed in South Vietnam, we know as a fact:

> **At a minimum, 132,596 gallons of Agent Orange and 77,215 gallons of Agent White were used in Khanh Hoa Province in 272 missions.

> **At a minimum, 104,815 gallons of Agent Orange and 2,075 gallons of Agent White were used on Ninh Thuan Province in at least 65 missions.

> **Cam Ranh was sprayed directly with a minimum of 1,373 gallons of Agent White concentrate and between 4,230 and 22,227 gallons of Agent Orange concentrate during at least 31 missions.

6. According to declassified records, an undetermined amount of the insecticide malathion was used on Cam Ranh and the rest of South Vietnam in over 1,300 missions. Likewise, an unknown amount of the insecticide DDT was used within Cam Ranh and on the people living there.

7. All the personnel stationed in Cam Ranh—as well as in many other areas—were exposed to a genuine cocktail of harmful cell-damaging substances. Chemicals that produced injuries and alterations within the immune, cardiovascular, endocrine, respiratory, and neurological systems of veterans' on a basic biological level whether presumed or not.

8. It is impossible for the DVA to presume "in good faith" or even to conceive of a less than 50 percent probability of these toxic compounds affecting veterans' without first actually studying and knowing the harmful genetic, cellular, and hormonal impacts they experienced in South Vietnam.

9. We know that during a debate on the Agent Orange Act of 1991, the late Senator John McCain gave this magnificent speech:

> Those who so nobly served our country and their families have been patient long enough. This should not be a political debate. They deserve answers. They deserve action. I believe

this is a fair and equitable approach to deal with the controversy that surrounds Agent Orange.

For too long, much too long, the government's response to Agent Orange has been based on opinion, *perhaps even politics*, but certainly not on facts. . . Our Vietnam veterans served our nation with dignity and honor. In spite of the risks, they answered our country's call to fight. It is time we settle the controversy over Agent Orange.

10. At a hearing in 1988, Senator John Kerry condemned the administration's efforts to escape responsibility for the injuries caused by Agent Orange and other pesticides. He made the following glorious statement:

To those who say, "We don't have enough evidence," I would ask how high does the body count have to go? How many Vietnam veterans have to die before we have "enough evidence" to start admitting the truth and compensating veterans?

This administration's rhetoric is out of touch with reality. They tell us that Vietnam veterans are "national heroes." But they continue to turn a deaf ear to Vietnam veterans who need help.

11. We know that even today the controversy over Agent Orange and all the other pesticides used in Vietnam continues to rage unabated and that the veterans who served in Vietnam are still dying from un-presumed illnesses, even with all the magnificent speeches given by our elected officials over the last half-century.

These are the facts as I have learned them over my years of research and investigation. They are quite different from the information being presented by the US government and the DVA. I would suppose that if the government keeps chanting "inadequate or insufficient evidence to determine an association" long enough and hard enough, folks may just come around to believing they are right and buy into their widely publicized illusions.

However, to do that, you must first forget about the mountain of evidence and studies presented just in this work. You would have to pretend that none of them existed. If you could do that, then you might very well ask, "Where is the proof that in-country veterans were really injured by the chemicals and conditions they were exposed to in Vietnam?" If you believe the government, then it does seem like evidence is missing. Nevertheless, you have to remember and think of the smoke and mirrors of classic misdirection.

The US government and its many agencies suspiciously agree—which is both curious and scary at the same time—that the evidence they have reviewed is inadequate or insufficient to determine an association between veterans' exposure to the multiple pesticides used during the war in Vietnam and their diverse cancers and illnesses. The DVA claims that they are basing their respective opinions of veterans' illnesses on available epidemiologic studies and data.

Now when our administrative agencies use expressions such as "inadequate or insufficient evidence to determine an association" or "insufficient consistency, quality, or statistical power to permit a conclusion regarding the presence or absence of an association," what are they telling—or maybe not telling—us? Their statements indeed sound authoritative and even convincing. So what can we learn from them?

Essentially, what they are telling us are their opinions. What is missing from their commanding declarations are facts. Not once do the DVA, NAM, and IOM in their authorized statements tell us that the chemicals we were exposed to weren't extraordinarily toxic, genetically damaging, or cell-altering substances. Instead, they use words and catchphrases that are designed to captivate or appease a wide-ranging audience by using the same statement(s) over and over again—in essence, a mantra that does not state what the evidence is, but sounds like it does.

Is There Really Missing Evidence?

Recall the keywords used by the US government and DVA: "available epidemiologic studies." Well, over the years, I must have read and reread that phrase countless times before what our governmental agencies were saying finally dawned on me. So what are the only epidemiologic studies available for the DVA to use for presumptive probability assessments?

Again, you are correct: civilian occupational research studies. The US government choosing to use civilian studies and data rather than to conduct research using in-country veterans is unquestionably quite significant and condemning. Using civilian data, as I have argued throughout this book, is the most significant cause of the theoretically missing evidence.

The evidence is not missing. It is being obstructed and suppressed. Although it is easy for me to make a statement like that, the question remains: Why has the US government refused to study veterans and opted to use and analyze unsuitable civilian study data instead? An excellent question like that deserves an answer. In reality, by just using civilian data and limiting the chemicals being studied to a select few, what our executives and leaders have so skillfully accomplished is to minimize the apparent illnesses caused by the various pesticides used during the war. Restricting the data to workplace/occupational incidents and just a few substances permits our governmental agencies to presuppose and agree only a small controlled number of diseases, immune system illnesses, and endocrine disturbances occurred in veterans who were heavily exposed to the several different toxic herbicides and insecticides.

As an added bonus, this executive manipulation and lack of studies allows the DVA to plausibly deny all the other illnesses and afflictions of in-country veterans by using the mantra "available epidemiologic studies." The missing evidence, veteran health studies, makes obtaining the most appropriate data to determine presumptive probability extrapolations challenging, if not impossible, for expert medical professionals, research groups, and even the DVA.

The mere fact that the government has not studied in-country veterans puts the DVA and their administrative agencies in an extremely indefensible position as well. The US government cannot escape the magnitude of its identifiable failure to act, which is just as unacceptable as denying the impacts and toxicity of its pesticides in the first place. The failure of our government to study Vietnam veterans actually constitutes a breach of duty, which could still give rise to legal responsibilities for that failure.

There is no way the DVA can honestly and in good faith make presumptive assessments of less than 50 percent for any of the far-reaching illnesses caused by our service in Vietnam without first knowing what happened to our cellular structure and health during the war.

The US government and DVA—not veterans—must prove, not just opine, that the chemicals and circumstances to which I and countless other veterans were exposed to and unprotected from in service:

1. Were not harmful to our health in any way;

2. Were not genetically damaging to our children;

3. Did not or could not physically damage or alter our cellular or genetic structure covertly; and

4. Did not or could not systemically injure our immune, endocrine, cardiovascular, respiratory, or neurological systems.

I am confident that, had our legislative and military leaders performed their legally and morally obligated precautionary tasks before and after the war instead of chanting meaningless words, we would know what happened to our health in Vietnam. We would know how our unprotected pesticide exposure and service affected us. We would know what that exposure did to the health and genetics of our children and precisely what the extensive list of chemicals still might do to the health of our grandchildren's children.

Largest Unstudied Environmental Disaster in the World

Dr. Jeanne Stellman, who wrote an article about Vietnam and Agent Orange in *Nature* magazine, acknowledged that "this is the largest unstudied environmental disaster in the world—except for natural disasters."[1] So is the Vietnam War really the largest unstudied disaster in our world today? The simple fact is that veterans from each and every one of our wars served the United States with honor under life-threatening conditions, and they all had their own set of unique challenges and burdens.

Nonetheless, in Vietnam, along with all the common difficulties present during any war, we also had a genuine quagmire of toxic chemicals to wade through. Our pesticide exposures were further complicated by the physical and mental stresses naturally created by any confrontational scenario, and Vietnam was no different. In point of fact, according to numerous medical studies, stress has many critical negative impacts on the human body, not the least of which is a lower ability to handle and compensate for multiple toxic chemical exposures.[2]

The issues created by the unstudied harmful impacts of all the very hazardous military pesticides we veterans were exposed to have simmered for decades. The effects of these uncountable dangerous chemical substances have been denied, manipulated, downplayed, and intentionally left unstudied by our government, military, and the DVA for almost half a century. Beginning in the 1970s and up through the 1980s, we ran the gamut, from the government interfering with and manipulating studies to their complete denial that exposure to tactical pesticides caused any alterations, illnesses, or disorders in veterans.

Thank God that despite all the government rhetoric and interference, numerous scientific and medical studies have been completed that confirm the adverse systemic medical and health problems that develop from exposure to the toxic chemicals investigated in this work. The real list of the cancers and illnesses triggered by the pesticides we veterans were exposed to has yet to be revealed. It has not even begun to be studied, let alone presumed by the DVA.

So regardless of the outcome of all the information-limiting by the US government, our legal battles, the DVA's defendable policies, or the skewed presumptive probability assessments, a significant reality must be recognized as a consequence of all the actions and inactions of our leaders. The simple truth is that since the end of the war, many cancers, illnesses, and disorders have claimed the lives and damaged the health of countless veterans. I can assure you with complete and utter confidence that the numerous complex and diverse cancers, illnesses, and maladies I and tens of thousands of other veterans—and even many of our children—are facing today are *not* being influenced or helped by governmental denials, controlling strategies, legal battles, or the DVA's "defendable policies."

The Name Recognition of Agent Orange

Last but not least is the fact that Agent Orange has become so recognizable. It has effectively dominated and hidden all the other highly toxic pesticides used during the war. Whether this was by chance or design—or just smoke and mirrors—is unknown. However, as a result, all the other pesticides have become lost in the shadow of Agent Orange.

Agent Orange will inevitably remain an everlasting symbol—an emblem, if you will—of all the highly toxic chemicals used during the Vietnam War. Nevertheless, the rest of the color-coded sunset pesticides and their associated deadly ingredients must also be acknowledged and studied; they must assume their rightful place in the history and tragedy of the Vietnam War.

After all is said and done, consider, if you will, that we are the greatest nation on earth. We have the preeminent scientific and medical communities in the world. Yet the DVA is trying to convince veterans that it was impossible for them to have conducted ongoing, definitive morbidity and mortality studies on how the Vietnam War and all its chemicals and circumstances affected the health of the military personnel who served there. I, for one, find that assertion quite doubtful, hypocritical, and disingenuous, especially given all the work that has been accomplished concerning Gulf War syndrome.

During my years of research, I have quite literally reviewed thousands of studies and documents. The vast majority of those records came to the same inescapable conclusions that I eventually did at the end of my investigation. Low-level exposures to just the various known chemicals recorded in this work will attack living organisms on an undetected hormonal, genetic, and cellular/molecular level, producing covert systemic damage and alterations to the immune, cardiovascular, endocrine, respiratory, and neurological systems of any human put in their path. What happened in Vietnam didn't stay in Vietnam. It came home with the men and women who served there, regardless of whether the US Government is willing to admit it or not.

End Notes - Chapter 19

1. Jeanne Stellman, "The Extent and Patterns of Usage of Agent Orange and Other Herbicides in Vietnam," *Nature* 422 (2003): 681–87.

2. Executive Summary: Vietnam Veterans, Chemical Exposures, Parkinson's Disease

Afterword

Regrettably, I cannot go back over the last half a century to get a *do-over* or to have the war conducted differently. I cannot force our legislative or military leaders to make better decisions. I cannot rewrite the unpleasant history of the Vietnam War, with all the numerous negative impacts that war had on me and every other soldier, marine, or sailor who served the United States in South Vietnam and in the blue waters of the surrounding ocean.

The very best I can do, almost a half century after the war, is to write, an account of our betrayal and describe our exposures to the toxic pesticides and abhorrent conditions of the Vietnam War, as best as I can. All in the sincere effort to correct the *present* so that what occurred in South Vietnam will never happen again to new generations of military personnel, their families and their children's grandchildren's children.

That brings me to the end of this investigational journey, and back to the question I asked in chapter one: Do you think the term *treacherously betrayed* is unduly harsh, correct, or overly lenient? Let me know!

https://www.ssdavw.com

Thank you and God Bless.

Postscript:
The End of One Book Is Just the Start of Another

As I was putting the finishing touches on my manuscript and reading through all the data and information again, it started me, once more, concentrating on what our government and the military-industrial chemical corporations were capable of creating in South Vietnam during the war. I began to spectacle, "on just how the United States got away with unleashing so many harmful pesticides during the war." Awkwardly, for me at least, even though I was there, the whole concept of what occurred in Vietnam is still quite perplexing and hard for me to fathom.

Still, based on my years of research, it appears that pesticide companies, our government, lumber companies, and large commercial agricultural groups, as well as many of our state and federal agencies, consider herbicides and insecticides essential for use in today's modern, industrialized world. Consequently, what occurred in Vietnam hasn't stayed in Vietnam. It has, over the intervening half century, continued to be ever so skillfully reproduced on a smaller scale in today's world as well. Like Vietnam, our government and chemical companies are still using the same classic trickery of smoke and mirrors for the express protection of harmful pesticides and their manufacturers. Health-damaging products that, by their very design and nature, are used to kill living entities.

In reality, it is quite easy to compare the modern commercialized pesticide products of today with those used in Vietnam because they contain some of the same hazardous organic chemicals. Furthermore, our modern pesticides are subject to the same government manipulation, and black hole secrecy as the ones used in Vietnam. I am pretty sure that most people would be able to look at all the chemicals we were exposed to and not suspect corruption or foul play. I certainly can't.

It is a scientific fact that pesticides will impact more than just the annoyances they were designed to target. They are unquestionably highly complex and toxic products by deliberate formulation. So if by intent pesticides are deadly, how do corporations get them approved? While the answer is easy, the toxic end-products being produced are far from simple.

In most cases, lobbyists working for large chemical conglomerates have an inordinate involvement in the drafting of our pesticide laws— albeit behind the scenes. The reason again is a straightforward one: Most of the legal text contained in our pesticide laws are of an exceptionally complex and technical nature. They are also exceedingly ambiguous. This vagueness in turn further helps pesticide industries—also by design—to allow their finished toxic products to be manufactured and approved for sale. As to the why, that is likewise very basic.

According to the EPA, worldwide pesticide sales at the manufacturer level totaled nearly $56 billion in 2012. By 2021, the global pesticide industry is expected to reach an estimated $81.1 billion in annual retail sales. Sounds pretty profitable, if you ask me. In stark contrast, tighter, more restrictive laws and regulations might very well lower that profitability substantially.

The primary support and strength of the chemical industry's protection in the United States originates with the testing afforded their many lucrative, but toxic formulations. It is the ambiguous testing that allows government agencies to approve or register harmful pesticides for use in the farming industry, the forestry service, and even on our lawns and in our homes.

As if inadequate vague testing of pesticides was not a serious enough matter, pesticide laws have also been laced with subtle loopholes that invite the industry to develop and sell even more injurious pesticides. Especially disturbing is the fact our government and the EPA will allow chemical companies to designate many detrimental substances and manufacturing contaminants found in their finished creations as "inert." Still, these hazardous compound are anything but inert; they are combinations of multidimensional ingredients, many of which include carcinogenic petroleum distillates and other dangerous substances not required by our laws to be disclosed by the manufacturer. Thus, allowing them to stay shrouded in secrecy.

We must all remember that as the U.S. government and chemical conglomerates talk with the presumed authority of science when they tell us not to worry—"You can shower in the stuff, and it wouldn't be a problem"—that is not the reality. When they promise more research—on our behalf—regarding the toxicity of the pesticides they let loose, they are not being honorable or genuine.

Although in-country Vietnam veterans are but one group of victims of widespread pesticide spraying and governmental cover-ups and misconduct, there are still many others. The extensive use of multiple scantily tested, harmful pesticides in today's world is also very problematic and can be toxicologically devastating.

Most pesticides being used today are very complex mixtures of various chemicals. Thus, when several different pesticides are used together in our environment, they are quite capable of creating an epidemic of complex harmful synergistic interactions—health injuring interfaces which will affect our children and us. As a result, veterans who were in Vietnam and their children, as well as you and your children, are all casualties of the same promised studies and investigations that will just never happen.

I cannot think of a better or more fitting way to conclude this book than with the following quote from an *unknown author*: In the history of the United States, there have been two important forces that have been willing to die for you:

Force one,

Is Jesus Christ and

Force two, is the American soldier.

One sacrificed Himself for your sins and soul.

The other sacrificed themselves for your freedom.

P

E

A

C

E

CPSIA information can be obtained
at www.ICGtesting.com
Printed in the USA
LVHW081016170120
643984LV00013B/640